Estimating Prediction Uncertainty from Geographical Information System Raster Processing: A User's Manual for the Raster Error Propagation Tool (REPTool)

By Jason J. Gurdak, Sharon L. Qi, and Michael L. Geisler

U.S. Geological Survey Center of Excellence for Geospatial Information Science (CEGIS)

Techniques and Methods 11–C3

U.S. Department of the Interior
U.S. Geological Survey

U.S. Department of the Interior
KEN SALAZAR, Secretary

U.S. Geological Survey
Suzette M. Kimball, Acting Director

U.S. Geological Survey, Reston, Virginia: 2009

For more information on the USGS—the Federal source for science about the Earth, its natural and living resources, natural hazards, and the environment, visit http://www.usgs.gov or call 1-888-ASK-USGS

For an overview of USGS information products, including maps, imagery, and publications, visit http://www.usgs.gov/pubprod

To order this and other USGS information products, visit http://store.usgs.gov

Suggested citation:
Gurdak, J.J., Qi, S.L., and Geisler, M.L., 2009, Estimating prediction uncertainty from geographical information system raster processing—A user's manual for the Raster Error Propagation Tool (REPTool): U.S. Geological Survey Techniques and Methods 11–C3, 71 p.

Contents

Figures

Tables

Conversion Factors

Inch/Pound to SI

Multiply	By	To obtain
Length		
inch (in.)	2.54	centimeter (cm)
inch (in.)	25.4	millimeter (mm)
foot (ft)	0.3048	meter (m)
mile (mi)	1.609	kilometer (km)
Area		
acre	4,047	square meter (m^2)
acre	0.004047	square kilometer (km^2)
square foot (ft^2)	929.0	square centimeter (cm^2)
square foot (ft^2)	0.09290	square meter (m^2)
square inch (in^2)	6.452	square centimeter (cm^2)
square mile (mi^2)	2.590	square kilometer (km^2)

Acronyms

CDF	cumulative distribution function
CEGIS	Center of Excellence for Geographic Information Science
DEM	Digital Elevation Model
GHz	gigahertz
GIS	Geographic Information System
GB	gigabytes
GUI	Graphical User Interface
LHS	Latin Hypercube Sampling
RAM	Random Access Memory
REPTool	Raster Error Propagation Tool
RMSE	Root Mean Squared Error
RVC	Relative Variance Contribution
USGS	U.S. Geological Survey

Preface

This report presents a computer program for estimating prediction uncertainty that is caused by error propagation during raster processing. The performance of this computer program has been tested in raster-processing models using environmental data; however, future applications of the program could reveal errors that were not detected in the test simulations. Users are requested to notify the U.S. Geological Survey (USGS) if errors are found in the report or in the computer program. Correspondence regarding the report should be sent to:

U.S. Geological Survey
Colorado Water Science Center
Mail Stop 415, Denver Federal Center
Lakewood, CO 80225

Although this program has been used by the USGS, no warranty, expressed or implied, is made by the USGS or the United States Government as to the accuracy and functioning of the program and related program material. Nor shall the fact of distribution constitute any such warranty, and no responsibility is assumed by the USGS in connection therewith.

The computer program documented in this report is available through the World Wide Web at the address:
http://co.water.usgs.gov/projects/REPtool/

Estimating Prediction Uncertainty from Geographical Information System Raster Processing: A User's Manual for the Raster Error Propagation Tool (REPTool)

By Jason J. Gurdak, Sharon L. Qi, and Michael L. Geisler

Abstract

The U.S. Geological Survey Raster Error Propagation Tool (REPTool) is a custom tool for use with the Environmental System Research Institute (ESRI) ArcGIS Desktop application to estimate error propagation and prediction uncertainty in raster processing operations and geospatial modeling. REPTool is designed to introduce concepts of error and uncertainty in geospatial data and modeling and provide users of ArcGIS Desktop a geoprocessing tool and methodology to consider how error affects geospatial model output. Similar to other geoprocessing tools available in ArcGIS Desktop, REPTool can be run from a dialog window, from the ArcMap command line, or from a Python script.

REPTool consists of public-domain, Python-based packages that implement Latin Hypercube Sampling within a probabilistic framework to track error propagation in geospatial models and quantitatively estimate the uncertainty of the model output. Users may specify error for each input raster or model coefficient represented in the geospatial model. The error for the input rasters may be specified as either spatially invariant or spatially variable across the spatial domain. Users may specify model output as a distribution of uncertainty for each raster cell. REPTool uses the Relative Variance Contribution method to quantify the relative error contribution from the two primary components in the geospatial model—errors in the model input data and coefficients of the model variables.

REPTool is appropriate for many types of geospatial processing operations, modeling applications, and related research questions, including applications that consider spatially invariant or spatially variable error in geospatial data as follows.

A. Analyses of error propagation, uncertainty, and sensitivity to understand and estimate:

- How error in geospatial model input propagate through Map Algebra expressions.

- The magnitude of spatially variable uncertainty of geospatial model predictions from error introduced as model input.

- The relations between spatially variable uncertainty of geospatial model predictions and error introduced from individual model inputs.

B. Given uncertainty in geospatial model predictions:

- Evaluate the range of probable predictions from geospatial models to environmental standards or regulatory limits.

- Determine the type and location of data needed to improve model prediction confidence.

- Assist best management strategies and decisionmaking under finite resources.

This report presents the theory and implementation of Latin Hypercube Sampling within the probabilistic framework for uncertainty analysis, capabilities and limitations of REPTool, detailed explanation of how to use REPTool, and developer-level documentation about Python-package architecture in REPTool. Additionally, example applications of REPTool are presented and illustrate that spatially variable prediction uncertainty of geospatial models can be quantified and used to reduce prediction uncertainty in future iterations of geospatial models.

Introduction

Error that is inherent to geospatial raster data can propagate through geospatial models that are used in geographic information systems (GIS) for many natural science and social science applications. The error propagation can result in substantial and spatially variable prediction uncertainty in model results. Consequently, prediction uncertainty has important implications for the use and interpretation of geospatial model results by scientists, environmental regulators, resource managers, elected officials, and the general public.

The propagation of input error from raster data and resulting prediction uncertainty of geospatial models, however, are rarely quantified or reported. Heuvelink (1999, p. 207) poses the question: "But exactly how large are the errors in the results of a spatial modeling operation, given the errors in the input to the operation?" Many GIS users may be aware of error propagation during geospatial modeling and question the confidence, or conversely, uncertainty, in the results of the operations but in practice rarely address or quantify this problem because of the lack of a universally available tool or methodology. As noted by Krivoruchko and Gotway (2005, p. 74), "GIS users need easily accessible tools for quantifying errors and assessing their impacts on resulting maps derived from geographical layers."

The ubiquitous use of raster data, the relative lack of awareness of error propagation by some GIS users, and the lack of available tools to address the inherent error propagation in geospatial models and resulting prediction uncertainty have prompted the following important questions that need to be addressed by scientists and other users of geospatial models. What function does the propagation of error from raster data have on prediction uncertainty of geospatial models? Can prediction uncertainty be quantified in geospatial models? Can prediction uncertainty be minimized in future geospatial models? Addressing these types of priority questions and a better understanding of the causes and effects of error propagation and uncertainty in model results are needed to establish the utility of data and models as decision support tools, to direct efforts toward improving data and models, and to identify alternative resource management strategies using the knowledge that model results from decision support tools are uncertain (Brown and Heuvelink, 2005).

To address these priority questions, the U.S. Geological Survey (USGS) Center of Excellence for Geospatial Information Science (CEGIS) supported development of the computer program called Raster Error Propagation Tool (REPTool). REPTool uses a probabilistic framework to identify the propagation of input error from raster data used during geospatial modeling and to quantify the prediction uncertainty that is associated with the geospatial model output. REPTool is a general error-propagation tool appropriate for many natural science and social science applications that use raster data and geospatial modeling.

Purpose and Scope

The purpose of this report is to describe the theory, implementation, structure, and operation of REPTool to estimate error propagation and prediction uncertainty during raster processing in a GIS. The report also includes examples of REPTool in geospatial modeling applications, which demonstrate that spatially variable prediction uncertainty of geospatial models can be quantified and an approach to reduce prediction uncertainty in future model runs using REPTool.

Overview of Raster Error Propagation Tool (REPTool)

The Raster Error Propagation Tool (REPTool) is a public domain, Python-based geoprocessing tool (computer program) that is designed for use with the Environmental System Research Institute (ESRI) ArcGIS Desktop application (ESRI; Redlands, Calif.) and may be accessed using a graphical-user interface (GUI). REPTool uses Latin Hypercube Sampling (LHS) (McKay and others, 1979) within a probabilistic framework to track error propagation from user-specified input raster data and a geospatial model to quantitatively estimate the prediction uncertainty of model output (fig. 1). The system overview of REPTool includes the input rasters, geospatial model, and model output (fig. 1). A user-specified error (spatially invariant or spatially variable error) for each input raster value is represented in the geospatial model as a user-specified distribution (Normal, Lognormal, or Uniform). Any coefficient used in the geospatial model may also have a user-specified error and the same distribution type as the input rasters. The LHS is used to efficiently propagate the error through each operator of the geospatial model and estimate the error propagation through to the final model output (fig. 1). REPTool enables the user to estimate a distribution of prediction uncertainty for each output raster cell (fig. 1). REPTool uses the Relative Variance Contribution (RVC) approach (van Horssen and others, 2002) to quantify the relative error contribution from the inherent errors in the model input data and coefficients of the model variables.

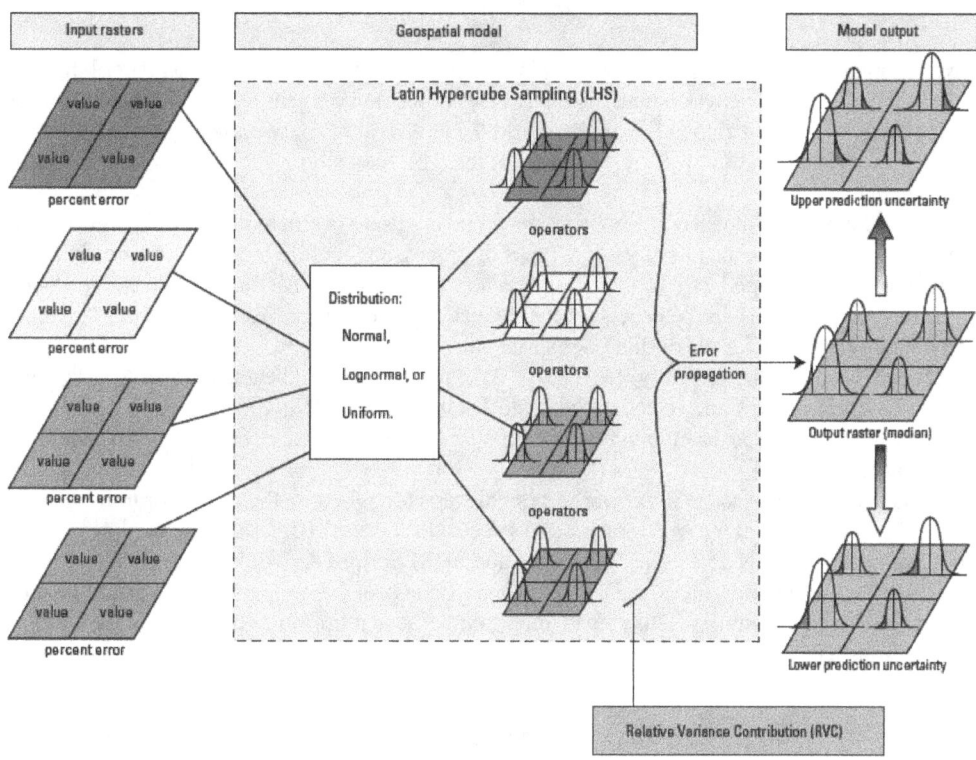

Figure 1.   System overview of Raster Error Propagation Tool (REPTool).

Getting Started

The following sections outline the basic information for obtaining, installing, and executing REPTool. Detailed information on using REPTool is provided in the section "**REPTool User's Guide**." Users that are unfamiliar with the topics of error propagation and uncertainty in geospatial operation and models will find additional background information in Brown and Heuvelink (2005), Foody and Atkinson (2002), Heuvelink (1998), Konstantin and Gotway (2005), and Zhang and Goodchild (2002).

System Requirements

- ArcGIS Desktop version 9.2 (ESRI; Redlands, Calif.) installed on a computer with Microsoft Windows operating system. ArcGIS is a suite of GIS software products produced by Environmental System Research Institute (ESRI; Redlands, Calif.) that allows users to create, edit, visualize, analyze, and output geographically referenced data. ArcGIS Desktop is used as a framework for REPTool because of its widespread availability and use among natural, social, and geographic-information scientists. The information in this report is presented under the assumption that the user has a working knowledge of the ArcMap, ArcToolbox, and Spatial Analyst extension and Map Algebra (Tomlin, 1990) concepts within ArcGIS Desktop.

- A Spatial Analyst license for ArcGIS Desktop version 9.2. ArcGIS Spatial Analyst is a set of tools for raster- and vector-based spatial modeling and analysis.

- Python version 2.4. Python is a dynamic, object-oriented programming language that has an open-source license. Python is not installed with REPTool and must be downloaded (*www.python.org/download/*) and installed independently from REPTool.

Installation and Execution of REPTool

The Raster Error Propagation Tool (REPTool) is a public domain, Python-based geoprocessing tool that may be downloaded and used within the ArcToolbox environment of ArcGIS Desktop (Environment System Research Institute, 2006; Redlands, Calif.). REPTool is available as a zipped file on the World Wide Web at *http://co.water.usgs.gov/projects/REPtool/* and *http://arcscripts.esri.com/*. The REPTool_v_1_0.zip file contains the REPTool source code, user interface, User's Manual, and other supporting documentation.

Once REPTool_v_1_0.zip is downloaded, it can be unzipped in any user-specified directory; however, because ArcGIS Desktop does not allow for spaces in pathnames for some types of processing, the file must be unzipped in a directory that does not have spaces in the path. The unzipped REPTool_v_1_0 directory contains the REPTool.tbx file and the "src" and "Docs" directories. The "src" directory contains all the necessary Python packages to run REPTool. The "Docs" directory contains support files for the REPTool interface and a file of this User's Manual.

After the files are unzipped into the directory REPTool_v_1_0, the toolbox can be added to an ArcMap session by right clicking within the Arc Toolbox window and selecting the "Add Toolbox" option (fig. 2).

In the Add Toolbox window (fig. 3), browse to the REPTool_v_1_0 directory, select REPTool.tbx, and click "Open" to add REPTool to the Arc Toolbox.

After the toolbox is added to the Arc Toolbox, one additional step is required by the user prior to running REPTool. The user right-clicks on the REPTool icon in the Arc Toolbox and selects "Properties" (fig. 4).

In the REPTool Properties window, select the Source tab, and in the Script File window browse to the …\REPTool_v_1_0\ src\main\ directory and select the main.py file (fig. 5). Click "Open" to accept the main.py as the Script File and click "OK" in the REPTool Properties to close that window. This step is important because it links the REPTool dialog window (GUI) with the Python source code that runs REPTool.

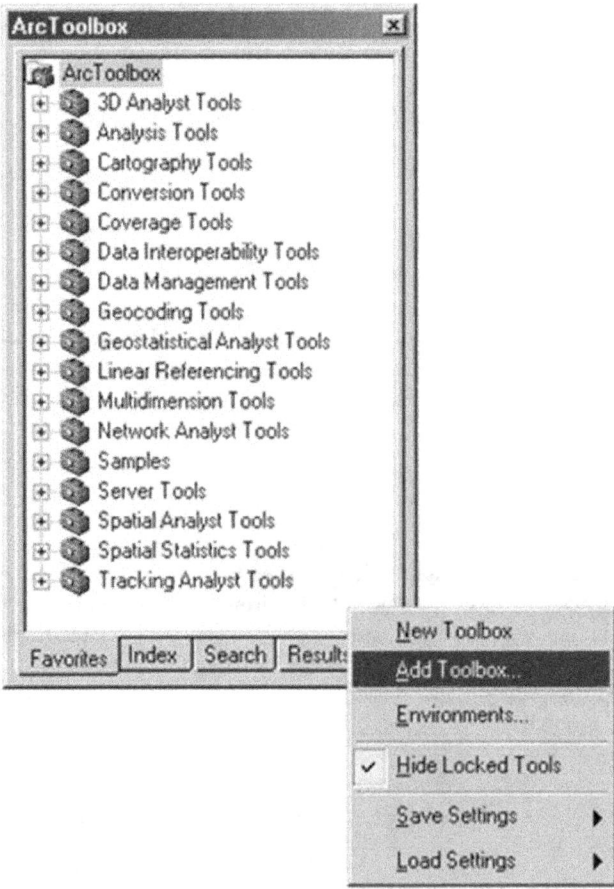

Figure 2.   Selection of the "Add Toolbox" option from the Arc Toolbox window.

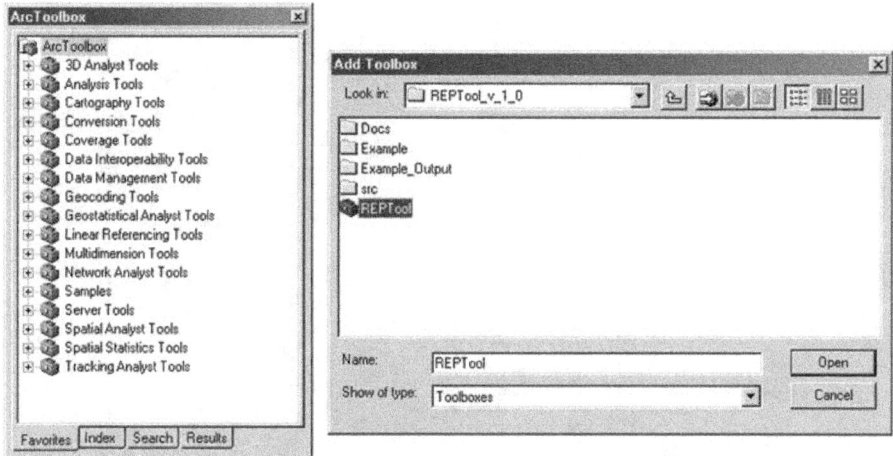

Figure 3.   The REPTool.tbx is selected from the REPTool_v_1_0 folder.

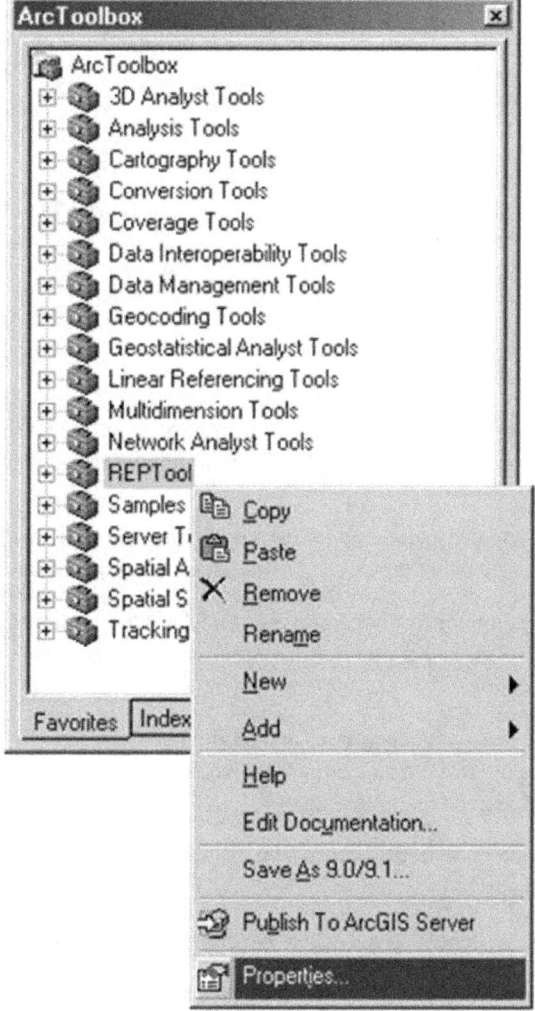

Figure 4.   The Properties option for REPTool is selected from the Arc Toolbox window.

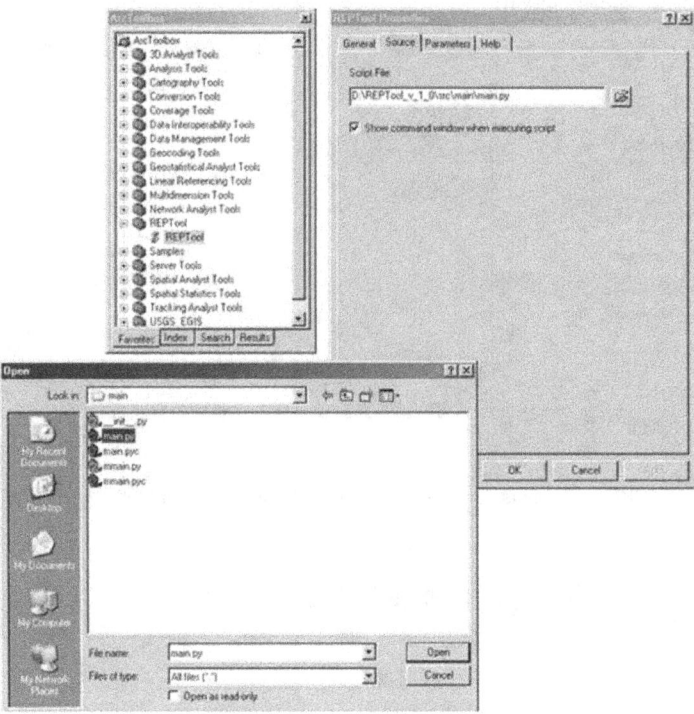

Figure 5.   Select main.py as the Script File in the Source Tab of the REPTool Properties window.

Program Capabilities and Characteristics

REPTool is a general error-propagation tool that quantifies uncertainty about model results by propagating user-specified error and is appropriate for geospatial operations, modeling applications, and research questions. REPTool has the following capabilities.

- REPTool can be integrated into any geospatial operation or model in an ArcGIS within the Map Algebra framework using raster data that are an estimate about the error(s) associated with those raster data.

- REPTool addresses error propagation for local operations in GIS [see "**Glossary**" for definition of local operation]. A local operation refers to those processes that produce an output value at single location (cell) that is dependent on the input-data value at that same location.

- REPTool integrates raster-based geospatial operations and modeling. Therefore, REPTool inherently includes all the capabilities and advantages of using raster data compared to vector data within a GIS. These capabilities and advantages are listed by Buckey (2008) and include the following.

  ○ The geographic location of each raster cell is implied by its position in the cell matrix and, thus, only the origin point coordinates are stored in the matrix.

  ○ Analysis of raster data is usually easier and quicker to perform than analysis of vector data.

  ○ Each cell in a raster data set stores one attribute, which makes raster data well suited for mathematical modeling and quantitative analysis.

  ○ Raster data facilitate and integrate discrete and continuous data.

- REPTool allows users to represent spatially invariant and spatially variable error that may be associated with input raster data used in a geospatial operation or model. The areal distribution of the spatially variable error must be provided by the user as a raster data set for input into REPTool.

- In addition to the results of the model equation, REPTool produces spatially variable raster maps of prediction uncertainty that are associated with the results.

- REPTool uses a probabilistic framework, which provides a number of advantages for error propagation and uncertainty analysis in geospatial models and includes the following.

   - REPTool uses LHS, which is a stratified Monte Carlo method and results in greater confidence in the sample distribution, fewer model simulations, and faster computation times than simple random Monte Carlo methods, which are necessary for application in GIS. Additional details of LHS and the advantages over Monte Carlo methods are described in the section "**Raster Processing, Error, and Uncertainty.**"

   - REPTool enables users to represent normal, lognormal, or uniform error distributions for raster values and coefficients in a Map Algebra framework of the geospatial model.

   - The probabilistic frameworks of REPTool offers users a number of applications, as first outlined by Krivoruchko and Gotway (2005), including:

      - Error propagation and uncertainty analysis to describe how the inherent error in rasters may affect the uncertainty of geospatial-model output.

      - Quantitative-risk analysis to describe the probability of an event occurring and the likely consequences should it occur.

      - Decisionmaking and resource-allocation analysis to evaluate "what if" scenarios and sensitivity assessments that may help the user in making the most informed decision.

Program Limitations

Although REPTool is a flexible and user-friendly program for estimating error propagation in geospatial models, the following list gives limitations of REPTool that should be considered by the user.

- There are some limitations and disadvantages of using raster data rather than vector data within a geospatial-model framework. REPTool is designed for application with raster data and therefore inherently includes all the limitations and disadvantages of using raster data rather than vector data within a geospatial-model framework. These limitations and disadvantages are listed by Buckey (2008) and include the following.

   - The raster cell size determines the resolution at which the data are represented, and some data may not be represented at the original resolution when converted to raster form.

   - Depending on the cell resolution, it may be difficult to represent detailed or spatially dense linear features in raster form.

   - The processing of raster attribute data may be challenging or cumbersome if large amounts of data are required in the geospatial model.

   - Although an advantage in most applications, raster data inherently represent only one attribute, which may be a limitation in some applications.

   - Because many spatial data are in vector form, the data must undergo a vector-to-raster conversion prior to being used in REPTool. In addition to increased processing requirements, a vector-to-raster conversion may introduce data integrity and error concerns.

- REPTool does not quantify the error that may be associated with the input raster data. The user of REPTool must specify the error associated with the input raster data and model coefficients. Previous research has focused on approaches for quantifying the six standard error components of GIS data, which are attribute accuracy, positional accuracy, lineage, logical consistency, completeness, and temporal accuracy (Heuvelink, 1998). Although the user is responsible for assigning the appropriate error for input raster data, REPTool may be most useful for analyzing the propagation of attribute accuracy because this error component is most important for raster data. The other five standard error components (positional accuracy, lineage, logical consistency, completeness, and temporal accuracy) are more important for vector data (Heuvelink, 1998). Some considerations for quantifying error of input raster data are provided in "**REPTool User's Guide.**"

- The LHS error-propagation framework that is implemented in REPTool follows the assumption originally outlined by McKay and others (1979) that input raster data, model coefficients, and corresponding error are statistically independent unless explicitly defined by the user in the REPTool model equation. Therefore, the current version of REPTool samples the distribution of the input rasters, model coefficients, and error as if they are statistically independent. The user should be aware that if correlation structure exists between input raster, model coefficients, and (or) error, the theoretical statistical properties of the output distribution may not be an accurate representation of the input correlation structure. Additional details are given in the section "**Latin Hypercube Sampling Method.**"

- REPTool does not explicitly account for heteroscedasticity or spatial autocorrelation but does enable the user to specify input rasters of spatially variable error that may have heteroscedasticity or spatial autocorrelation characteristics.

- REPTool is limited in the number of unique variables and coefficients that may be used in a geospatial model equation. REPTool allows for 100 unique variables (var00 to var99) and 100 unique coefficients (c00 to c99) within the geospatial model.

Raster Processing, Error, and Uncertainty

The scope of raster processing ranges from simple binary operations (that is, raster A–raster B) to complex computational models. Often, raster processing is used to derive new geospatial data for either stand-alone applications or for use in more complex natural science or social science models. The use of raster processing to derive new raster data from existing rasters is one of the most common types of manipulations and most powerful capabilities of GIS (Heuvelink, 1999). The new rasters commonly are derived using various operations and functions within the Map Algebra framework. Additionally, rasters are particularly useful for quantitative-spatial modeling. The data contained in rasters are frequently the starting point for deriving secondary data and input used in models for decisionmaking (Heuvelink and others, 1989). If information about error is not recorded in the metadata of the primary or intermediate raster products, then the accuracy of the final product will be uncertain.

The popularity and far-reaching application of raster processing has created a substantial challenge for scientists and GIS practitioners—endemic uncertainty associated with geospatial model predictions because of propagation of inherent error during raster processing (Mowrer and Congalton, 2000). More importantly, many GIS practitioners may be aware of error propagation during raster processing, but in practice rarely address or quantify the error propagation because of the lack of a universally available tool or methodology.

This challenge stems from the unavoidable and inherent error associated with geospatial (raster) data in GIS as imperfect representations of the real world (Zhang and Goodchild, 2002; Hunsaker and others, 2001; Burrough and McDonnell, 1998). The two main types of error in GIS include the source error that exists in geospatial data used as input and the propagation of error through operations performed on these data (Heuvelink, 1998). The source error of geospatial data is defined by the difference between reality and the representation of reality in the geospatial data and generally is a function of the accuracy and precision of the geospatial data (Mowrer and Congalton, 2000; Heuvelink, 1998; Heuvelink and others, 1989). Accuracy of geospatial data refers to the closeness of represented measurements or computations to their "true" or accepted values, and precision refers to the level of measurement and exactness of descriptions reported in the geospatial data (Gottsegen and others, 1999).

The many different sources of error in geospatial data are discussed in detail by Veregin (1999), Burrough and McDonnell (1998), Heuvelink (1998, 1999), and Burrough (1986). Heuvelink (1998) lists the six standard-error components of geospatial data as attribute accuracy, positional accuracy, lineage, logical consistency, completeness, and temporal accuracy. Although it is beyond the scope of this report to describe all sources of error and methods to estimate those errors, a general overview is presented in "Theory of Error Propagation and Uncertainty."

Theory of Error Propagation and Uncertainty

The theoretical basis of error propagation and the related computations in REPTool is presented in this section. Although recent research efforts have improved quantification methods for many types of source error associated with geospatial data in GIS, no generally accepted theory exists for handling error propagation and uncertainty in GIS (Heuvelink, 1998). The error-propagation theory that is implemented in REPTool and presented in this report largely follows a stochastic error modeling framework that is previously described by Heuvelink (1998, 1999) and Heuvelink and others (1989).

Error propagation occurs because the output from a raster process or geospatial operation is a function of the input raster data sets, which have inherent source error that automatically affects the computed results (Heuvelink, 1998). The cause of error propagation is generally more complex because source error is not the only type of error that propagates through raster processing. Many raster processing applications use simple and complex computational methods (Gurdak and Qi, 2006; Qi and Gurdak, 2006; Gurdak, 2008) with coefficients or model structure that are subject to estimation error (van Horssen and others, 2002). Therefore, the uncertainty of results from raster processing is a function of error propagation from both source error of geospatial data and the computational error introduced by the geospatial model.

The result of error propagation is uncertainty in the model output. The uncertainty may create model output that is not sufficiently reliable for correct interpretation (Heuvelink, 1998). Moreover, the error-propagation and prediction uncertainty is further compounded when the output from one geoprocessing operation or model is used as input to a subsequent geoprocessing operation or model (Heuvelink, 1998). Tracking error propagation through various steps of a geoprocessing model may provide quantitative estimates of the uncertainty that is associated with the output and thus provide valuable information for those that subsequently use or interpret the output.

Quantitative Error Model

As defined by Heuvelink (1998, 1999), error is the difference between reality and the representation of reality that is expressed by the geospatial (raster) data. Error may be the result of human mistakes in measurement, in translating reality into geospatial data, or from natural spatiotemporal variation. Following this definition, error, $v(x)$, can be quantified as

$$v(x) = a(x) - b(x) \tag{1}$$

where

$v(x)$ is the error,
$a(x)$ is the true value at some location x, and
$b(x)$ is the representation of the true value by the geospatial data at location x.

Although $b(x)$ is known exactly because it is the attribute value in the geospatial data, the definition expressed in equation (1) assumes that the exact values of $a(x)$, and thus $v(x)$, are unknown. As described previously in this report, error is inherent to geospatial data and thus creates the logical inequality that $a(x) \neq b(x)$. If $a(x)$ were known exactly, GIS users would simply assign $b(x) = a(x)$ in the geospatial data and thus have no error term, $v(x)$, as in equation 1.

Consequently, the error term, $v(x)$ in equation 1, is never known exactly. However, using a stochastic method, the distribution of possible $v(x)$ values can be estimated and represented as a random variable $V(x)$ (Note—the capital notation denotes a random field or function from a deterministic variable). The stochastic method enables a conceptualization that the uncertainty about $a(x)$ (as expressed in the error, $v(x)$) can be represented as a random mechanism, $A(x)$, even though $a(x)$ has one deterministic value in reality (Heuvelink, 1998, 1999). Thus, equation (1) becomes

$$V(x) = A(x) - b(x) \tag{2}$$

where

$V(x)$ is the random error field (or function) representing the distribution that surrounds the deterministic value of $v(x)$,
$A(x)$ is the random field (or function) representing the distribution that surrounds the deterministic value of $a(x)$, and
$b(x)$ is the representation of the true value by the geospatial data at location x.

Rearranging equation (2) results in the quantitative stochastic error model:

$$A(x) = b(x) + V(x) \tag{3}$$

where

$A(x)$, $b(x)$, and $V(x)$ are as defined previously in equation (2).

The uncertain and variable nature of $v(x)$ at any one location x, which is expressed as the random error field $V(x)$ (eq. 3), allows for $V(x)$ to be quantified as a probability distribution. The mean and standard deviation of $V(x)$ are denoted by the expected value, or mean value, $E[V(x)] = \mu(x)$ and standard deviation $[\sigma(V(x)] = \sigma(x)$. The $\mu(x)$ is the systematic error or bias and represents how much $b(x)$ differs from $A(x)$, and the $\sigma(x)$ is the nonsystematic error and represents the random component of $V(x)$ (Heuvelink, 1998). The $\mu(x)$ and $\sigma(x)$ may be used to define the cumulative distribution function (CDF) of $V(x)$, $F_{v(x)}$ ($\cdot$), or the probability distribution function of $V(x)$, $f_{v(x)}(\cdot)$. Although it is frequently assumed in error analysis that $V(x)$ follows a normal (Gaussian) distribution, it may be more appropriate in some applications for $V(x)$ to follow another distribution type, such as a lognormal or uniform distribution (Heuvelink, 1998). The motivation behind the use of the normal distribution for $V(x)$ is the central limit theorem, which generally states that the average of a large number of random variables yields a normal distribution regardless of the distribution of the individual random variables (Heuvelink, 1998). The use of the lognormal distribution for $V(x)$ may be more appropriate in natural-science applications because many variables observed in the natural sciences follow a lognormal distribution (Heuvelink, 1998). A uniform distribution may be more appropriate distribution for $V(x)$ when only a range for $v(x)$ is estimated (Sanchez and Blower, 1997; Swartzman and Kaluzny, 1987).

Assigning Error

Raster data sets are rarely free of error, $v(x)$, (see eq. 1). However, GIS users face inherent challenges in estimating error and assigning $V(x)$ to the quantitative stochastic error model (eq. 3). In practice, the challenge of estimating error often is the result of the users, being many steps removed from the producers of raster data sets, and the metadata may inadequately document the error sources (Longley and others, 1999). Many methods for estimating raster error are described in the literature, including those by Heuvelink (1998, 1999), Krivoruchko and Gotway (2005), Longley and others (1999), Mowrer and Congalton (2000), Veregin (1999), and Zhang and Goodchild (2002).

In general, the random error field, $V(x)$, can be assumed to be spatially invariant or spatially variable on the spatial domain, D, of interest. Heuvelink (1999) suggests assumptions of spatially invariant error may be appropriate if observed error at test points of D are available and support that assumption. For example, a common type of geospatial data found in many natural-science applications results from geostatistical interpolation techniques of point observations. The $\sigma(x)$ of the resulting inter-polated geospatial data may be assumed to be spatially invariant and thus estimated by the root mean squared error (RMSE) (Heuvelink, 1999; Veregin, 1999). As an example, the error inherent to the USGS 30-meter digital elevation model (DEM) of the conterminous United States (U.S. Geological Survey, 1997) is reported in terms of the RMSE, but the error values likely vary spatially depending on the data and methods used to create the DEM (Bishop and others, 2006). Another possible approach for assigning spatially invariant error may involve a sensitivity analysis to determine the uncertainty of the model results to various user-assumed error percentage values (Helton and Davis, 2003). For example, Emmi and Horton (1995) evaluate the sensitivity of property damage and casualty predictions from a seismic risk assuming a 5-percent (increase or decrease) error in ground-shaking intensity values, which are input data to the seismic-risk model. The 5-percent error was propagated through the seismic-risk model using a stochastic method (Monte Carlo methods, see "**Latin Hypercube Sampling Method**") and indicate an 11-percent lower to 15-percent greater predicted damage for residential structures than model predictions not including the 5-percent error in ground-shaking intensity values (Emmi and Horton, 1995). The sensitivity-analysis example by Emmi and Horton (1995) illustrates that user-assigned error may be somewhat arbitrary to evaluate the sensitivity of models and "what if" scenarios of model predictions. For example, additional user-specified percent-error values of ground-shaking intensity, such as 10, 15, and 20 percent, could also have been used in the sensitivity-analysis example.

For some geospatial models, the assumption of spatially invariant error may not be appropriate. Heuvelink (1996, 1999) suggests that spatially variable error may be estimated from the spatial variability of the attributes of interest and the mapping procedure used to create the raster of interest, such as kriging. For example, observed error at test points of D may indicate systematic, rather than random, differences in error values among various regions of D. Observed error at test points of D may indicate that the variance of error differs (increases or decreases) with changes (increases or decreases) in magnitude of the spatial attribute represented by the raster, which is called heteroscedasticity. Similarly, the systematic differences in error values among various regions of D may indicate spatial autocorrelation between spatial attributes and corresponding error values (see Heuvelink [1999] for additional details). REPTool does not explicitly account for heteroscedasticity or spatial autocorrelation but does enable the user to specify input rasters of spatially variable error that may have heteroscedasticity or spatial autocorrelation characteristics.

Application of Quantitative Error Model and Error Propagation in GIS

The following section describes an application of the quantitative stochastic error model and error-propagation theory to a geospatial analysis that was first described by Heuvelink (1996, 1999) and Heuvelink and others (1989). REPTool assumes that the geospatial analysis is composed of only local GIS operators. REPTool does not track the movement of error through other types of GIS operations (such as neighborhood, flow-directed, or zonal operations).

Local operations in GIS produce an output value at a location, $P(x)$, that depends on the input, $A_i(x)$, at that same location, x, where each $A_i(x)$ location is therefore treated as spatially independent and is expressed as:

$$P(x) = g\left[A_1(x), A_2(x), A_3(x), ..., A_n(x) \right] \tag{4}$$

where

> $P(x)$ is the output of the geospatial operation or model, $g(x)$, in GIS at location x,
> $g(x)$ is the geospatial operation or model at location x, and
> $A_i(x)$ is the value at location x in each input raster, where i differentiates the inputs.

In practice, $g(x)$ is typically applied to D using a Map Algebra expression. Under the assumption of local operations, $g(x)$ is computed independently for all cells in D, and any spatial contiguity in the output values of $P(x)$ is the result of spatial contiguity of input attribute values from $A_i(x)$ (Heuvelink and others, 1989). The $g(x)$ may take the form of a predefined local operation from the Spatial Analyst extension in ArcGIS Desktop or user-specified empirically derived operations or models. Uncertainty of empirically derived GIS models have been evaluated recently in a number of applications, including modeling of groundwater vulnerability to nonpoint-source contamination (Gurdak and Qi, 2006; Gurdak and others, 2007; Gurdak, 2008); pedotransfer functions to predict soil hydraulic properties (Finke and others, 1996); regional vegetation models (van Horssen and others, 2002); landscape classifications (Canters and others, 2002); wildfire behavior modeling (Bachman and Allgöwer, 2002); and general modeling of environmental variables (Heuvelink and others, 2007) [Note—$P(x)$ frequently is used as input to complex models executed outside a GIS, which is illustrated with an example application of REPTool herein].

Empirically derived GIS models commonly contain model coefficients. For example, in a logistic-regression model, $g(x)$ may take the form of:

$$g(x) = \frac{e^{b_0 + \left[b_1 * A_1(x)\right] + \left[b_2 * A_2(x)\right] + ... + \left[b_n * A_n(x)\right]}}{1 + e^{b_0 + \left[b_1 * A_1(x)\right] + \left[b_2 * A_2(x)\right] + ... + \left[b_n * A_n(x)\right]}} \tag{5}$$

where

> $g(x)$ is the geospatial operation (logistic-regression model in this example) at location x,
> e is the base of the natural logarithm,
> b_i are the coefficients of the logistic-regression model, and
> $A_i(x)$ are the values at location x from each input raster of the logistic-regression model.

The b_i are rarely known exactly for all locations, x, and therefore the values of b_i are often estimated in most modeling approaches (Heuvelink, 1998). For example, Gurdak and Qi (2006) and Gurdak and others (2007) use the Wald 95-percent confidence limit on the logistic regression model coefficients, b_i, as an estimate of error for error-propagation analysis of a logistic regression-based groundwater vulnerability model of the High Plains aquifer. Therefore, error-propagation analysis may consider error that is associated with the raster data, $A_i(x)$, and error that is associated with the model coefficients, b_i, in the geospatial operation, $g(x)$.

Consequently, the general objective of evaluating error propagation is to estimate the error in the output of the geospatial operation or model, $P(x)$, given that the operation, $g(x)$, and input rasters, $A_i(x)$, contain error (Heuvelink, 1999). Similar to $V(x)$ and $A(x)$, the output raster, $P(x)$, can be considered as a random field (or function) with a distribution that is described by a mean, $\zeta(x)$, and standard deviation, $\tau(x)$. Therefore, the standard deviation, $\tau(x)$, or other measures of spread such as variance, $\tau^2(x)$, provide meaningful information about the effects of error propagation on the magnitude of uncertainty of $P(x)$ (Brown and Heuvelink, 2005). In order to represent the uncertainty of $P(x)$ as a statistical distribution function, the domain of the probability model must be specified under a set of assumptions that are described in detail by Brown and Heuvelink (2005). The integrals that define the domain of probability, typically expressed as a CDF, are not amenable to a closed-form evaluation and solution (Helton and Davis, 2003). Therefore, approximation procedures, such as Monte Carlo methods, are frequently used to estimate $\zeta(x)$ and $\tau(x)$ from the CDF of uncertainty surrounding $P(x)$ (Heuvelink, 1999). Although simple random Monte Carlo sampling method is frequently used in error propagation analyses, the Latin Hypercube Sampling method provides a number of advantages that are more appropriately suited for application in GIS and are described next.

Latin Hypercube Sampling Method

Monte Carlo methods have been successfully applied to error-propagation and uncertainty studies of GIS models (Brown and Heuvelink, 2007; Bishop and others, 2006; van Horssen and others, 2002; Sklar and Hunsaker, 2001; Phillips and Marks, 1996; Fisher, 1991). Latin Hypercube Sampling (LHS) (McKay and others, 1979) is a stratified sampling variation on the simple random Monte Carlo sampling method and provides an efficient method for GIS-based error-propagation and uncertainty analyses (Gurdak and Qi, 2006; Gurdak and others, 2007; Gurdak, 2008). To illustrate the efficiency and advantages of LHS over simple random Monte Carlo sampling in GIS applications, an explanation of the Monte Carlo method is first presented.

The simple random Monte Carlo sampling method involves generating repeated and random sample observations from a probability distribution of an uncertain variable. The sample observations often are used to characterize the uncertainty surrounding that variable and may be used as input to a similarly repeated number of computer simulations that result in a large number of possible output realizations. Specific to the "**Theory of Error Propagation and Uncertainty**" described previously, Monte Carlo methods may be used to calculate the result of $g(x)$ for a large number of times, N, with input values for b_i and $A_i(x)$ that are randomly sampled from their respective distribution functions. The result of the large N is a random sample of $P(x)$ from which statistic parameters, such as $\zeta(x)$, $\tau(x)$, and $\tau^2(x)$ can be estimated. The N must be sufficiently large that the random sample, or set of realizations, accurately represent $P(x)$. The accuracy of estimates from simple random Monte Carlo sampling is inversely related to the square root of the number of realizations and, as noted by Heuvelink and Burrough (1993), indicates that the accuracy of simple random Monte Carlo sampling slowly improves as N increases. For example, to double the accuracy of a particular set of simple random Monte Carlo realizations, four times as many N are needed (Brown and Heuvelink, 2005). Heuvelink (1998) reports that many practical applications of simple random Monte Carlo sampling use $N = 50$ to $2,000$. Therefore, the main disadvantage of simple random Monte Carlo sampling is the computational time that is required to calculate $g(x)$ at N times for each b_i and $A_i(x)$ when the number of raster cells in typical applications of $A_i(x)$ may exceed many tens to hundreds of thousands, depending on the cell size and resolution of the data. Although some studies have implemented simple random Monte Carlo sampling in GIS models using $N = 10$ or 20, Goodchild and others (1992) note that using small N values, such as 10 or 20, during simple random Monte Carlo sampling is not sufficient to obtain accurate estimates of $\zeta(x)$.

In contrast, LHS implements a type of stratified Monte Carlo method rather than a purely random sample with traditional Monte Carlo methods. Using a general notation scheme (Swiler and others, 2004; Wyss and Jorgensen, 1998), LHS selects n different values from each of uncertain k variables $X_1, X_2, ...,X_k$ in the following steps:

(i)   The range of each variable [b_i and $A_i(x)$ in the specific case of equations 4 and 5] that will be sampled is divided into n nonoverlapping intervals on the basis of equal probability (that is, such that the probability of falling in any of the intervals is $1/n$).

(ii)   One value is randomly sampled (selected) from each interval with respect to the CDF in the interval.

(iii) The n values that are randomly sampled from each interval are randomly paired (without replacement) for each combination of X_k. For example, the n values obtained for X_1 are paired in a random manner (that is, equally likely combinations) with the n values of X_2. The n pairs for X_1 and X_2 are combined in a random manner with the n values for X_3, and so on, until n k-tuplets are created (Swiler and Wyss, 2004; Wyss and Jorgensen, 1998).

(iv)   The result of this random pairing is n combinations of k variables, which represent an $(n \cdot k)$ matrix of input data where the ith row contains specific values of each of the k input variables to be used on the ith model simulation.

The main advantage of LHS over simple random Monte Carlo sampling is that the stratified sampling of the n nonoverlapping intervals in LHS requires fewer samples to accurately describe the CDF of interest. The stratified sampling technique of LHS produces a distribution of samples that more closely corresponds to the input probability distribution (McKay and others, 1979).

To illustrate the difference in sampling techniques between LHS and simple random Monte Carlo methods, consider an example of LHS where $n = 10$ and input variables X_1 and X_2 are both normally distributed (Wyss and Jorgensen, 1998). The CDF for X_1 and X_2 are described by μ values equal to 0 and 5 and σ values equal to 1 and 1, respectively. The CDFs and corresponding 10 nonoverlapping intervals, each defined by quantiles having an equal 0.1 probability, for X_1 and X_2 are shown in figure 6.

Once the nonoverlapping intervals of equal probability are identified for each variable (fig. 6), the next LHS step is the random sampling of specific values from the actual distribution domain (x-axis, fig. 6) for X_1 and X_2 in each of the $n=10$ respective intervals. In practice, however, the random sampling is done relative to the CDF (y-axis, fig. 6) rather than the actual distribution values (x-axis, fig. 6) because the nonoverlapping intervals are defined by the 10 quantiles of equal probability that are expressed as cumulative-probability units from 0 to 1 (y-axis, fig. 6). For this example, the randomly selected cumulative probability, p, from each of the 10 nonoverlapping intervals is listed in table 1. The actual distribution values, x, (x-axis, fig. 6) are calculated by inverting the CDF (Note—REPTool uses an algorithm defined by Acklam (2004) to invert the normal cumulative distributions. The equations defining the inverse cumulative distributions for normal, lognormal, and uniform

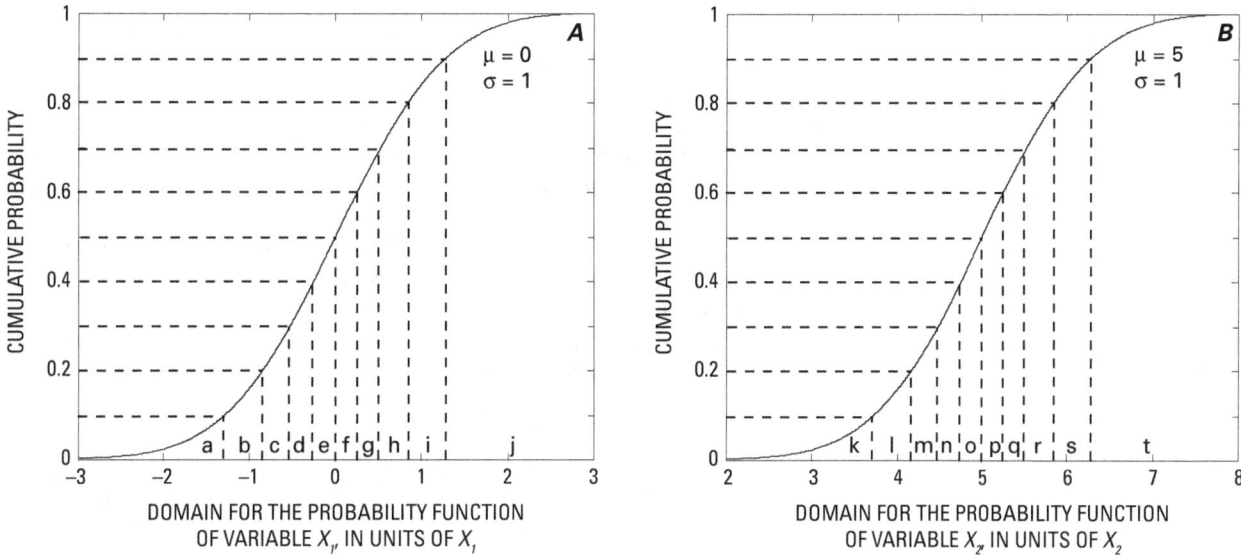

Figure 6. As an example of the stratified Latin Hypercube Sampling approach, n=10 nonoverlapping intervals having equal probability of 10 percent are shown on cumulative distribution functions (CDF) for (*A*) X_1 as intervals a–j, and for (*B*) X_2 as intervals k–t. The mean (μ) and standard deviations (σ) for each CDF are shown in the figure.

Table 1. Summary of randomly selected values for Latin Hypercube Sampling example.

Interval for X_1	Cumulative probabilities, p, within the interval	Corresponding actual distribution values, x	Interval for X_2	Cumulative probabilities, p, within the interval	Corresponding actual distribution values, x
a	0.04	−1.9	k	0.09	3.6
b	0.14	−1.1	l	0.19	4.1
c	0.22	−0.7	m	0.26	4.4
d	0.37	−0.4	n	0.35	4.6
e	0.43	−0.3	o	0.46	4.9
f	0.55	0.3	p	0.54	5.1
g	0.64	0.35	q	0.65	5.4
h	0.79	0.65	r	0.71	5.6
i	0.84	1	s	0.88	6.1
j	0.95	1.45	t	0.99	7.1

distributions are described in Appendix 1 of this report). The corresponding distribution values, x, for each randomly selected probability, p, are in table 1 and figure 7.

Next, the 10 values from X_1 are randomly paired (without replacement) with the 10 values from X_2 as two dimensional input vectors for each realization of the model equations (table 2). Assuming this demonstration example uses the model equation $Y = X_1 + X_2$, then the calculated Y values are summations of the randomly paired 10 values from X_1 and X_2 (table 2). For example, for model simulation number 1, the input vector is the first permutation vector (0.3, 4.4) from randomly paired sets of X_1 and X_2 and the model result, Y, is 4.7 (table 2). The input vectors for the second and subsequent simulations are constructed in a similar manner. The sorted model results in table 2 illustrate the distribution of uncertainty surrounding Y given error distributions of X_1 and X_2.

It is important to note that the current version of REPTool uses the McKay and others (1979) implementation of LHS that independently samples and randomly pairs the uncertain k variables, as demonstrated in table 2. Even though k variables are sampled independently and paired randomly, the sample correlation coefficients of the n pairs of variables may not equal zero due to sampling fluctuations (Swiler and Wyss, 2004). Furthermore, Iman and Conover (1982) state that if a correlation structure exists among the k input variables, but the actual LHS assumes independence among k variables, then the theoretical statistical

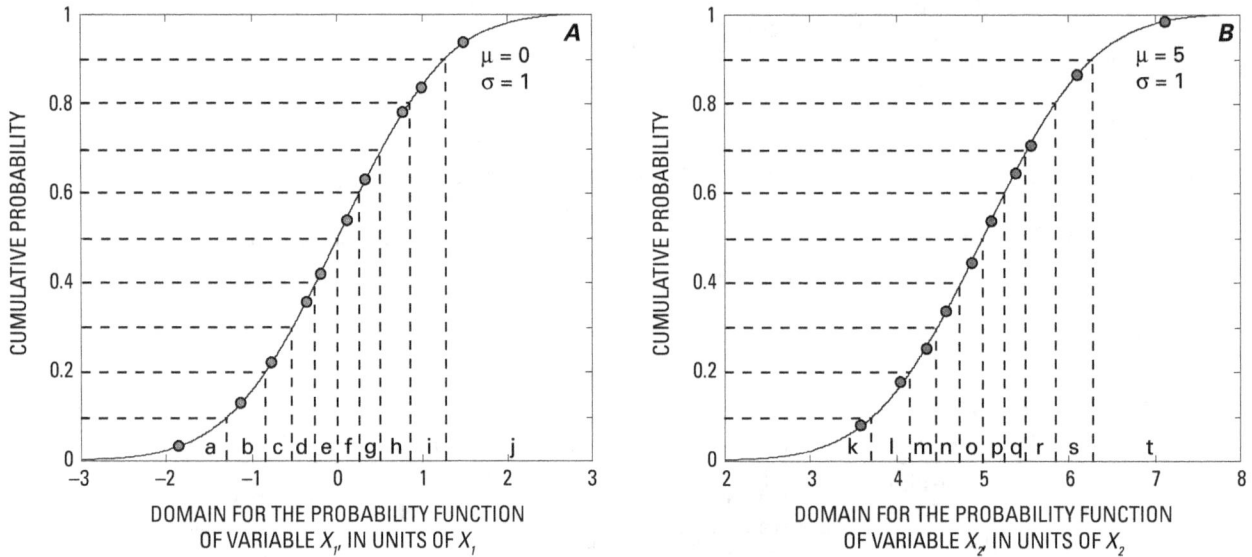

Figure 7. Randomly selected values using Latin Hypercube Sampling (LHS) for (A) X_1 and (B) X_2.

Table 2. Randomly paired values and model results from the Latin Hypercube Sampling example.

Model simulation number	X_1, randomly paired set	X_2, randomly paired set	Model result, Y ($Y=X_1 + X_2$)	Rank of model result, Y
1	0.3	4.4	4.7	5
2	0.65	5.1	5.75	8
3	−0.4	4.9	4.5	4
4	−1.9	5.6	3.7	2
5	1.45	6.1	7.55	10
6	−0.3	5.4	5.1	6.5
7	−0.7	7.1	6.4	9
8	0.35	3.6	3.95	3
9	−1.1	4.6	3.5	1
10	1	4.1	5.1	6.5

properties of the *P(x)* may not adequately describe the input correlation structure. REPTool users may use methods in the Geostatistical Analyst extension of ArcGIS Desktop to determine the extent and structure of any spatial correlation between data sets. For example, the cross-covariance cloud can be used to detect the extent of spatial correlation between input data sets, and the search direction tool can be used to indicate directional structure or asymmetry of the correlation between data sets.

An example of simple random Monte Carlo sampling (*N*=10) for X_1 and X_2 (fig. 8) illustrates that observation samples from Monte Carlo sampling typically do not characterize the CDF as well as LHS (fig. 7)—particularly the tails of distribution functions. To improve the characterization of the CDF for X_1 and X_2 using Monte Carlo methods, *N* must be increased. In general, for the same number of samples, LHS produces a more unbiased estimate of the mean and a smaller variance as compared to simple random Monte Carlo sampling. The smaller variance from LHS translates into a greater confidence, fewer model simulations, and faster computation times necessary for use within ArcGIS (Gurdak and Qi, 2006). This is especially beneficial for complex geospatial model simulations because executing enough simulations to properly represent the input distribution may be impractical using Monte Carlo methods. Although no absolute rule exists for *n* in LHS, McKay and others (1979) indicate that an *n* equal to twice the number of uncertain *k* variables may provide an adequate balance between accuracy and computational costs. Similarly, other studies indicate that satisfactory results are produced using *n* from 1.3 to 3 times the number of uncertain *k* parameters (Iman and Helton, 1988; Manache and Melching, 2004).

REPTool User's Guide

Similar to other geoprocessing tools available in ArcGIS Desktop, REPTool can be run from a dialog window (GUI) (fig. 9), from the ArcMap command line, or from a Python script. The "**REPTool User's Guide**" section describes how to use REPTool from a dialog window (GUI). The command-line syntax and scripting usage of REPTool are described in Appendix 2 of this report and in the REPTool Help page. The main REPTool Help page is accessed by clicking on "Show Help>>" at the bottom of the REPTool dialog window (fig. 9) and then by clicking on the "Help" button at the top-right of the expanded dialog help.

The REPTool dialog window (GUI) (fig. 9) consists of a sequence of entry fields that enable the user to specify input and output information for the Map Algebra model equation. The required fields include Input rasters, the Distribution type, and the Output workspace. All other fields are optional. The Advanced Parameters of REPTool (fig. 9) can be used to input spatially variable error and calculate the Relative Variance Contribution (RVC) of components in the model equation. Each user-entry field, from the top of the dialog window to the bottom, is described in the subsequent sections of the report. The user is encouraged to pay particular attention to the "**Model Equation**" section of this report because REPTool requires specific syntax for the input Map Algebra model equation that is somewhat different than the standard syntax for other ArcGIS Desktop applications, such as the Raster Calculator in Spatial Analyst.

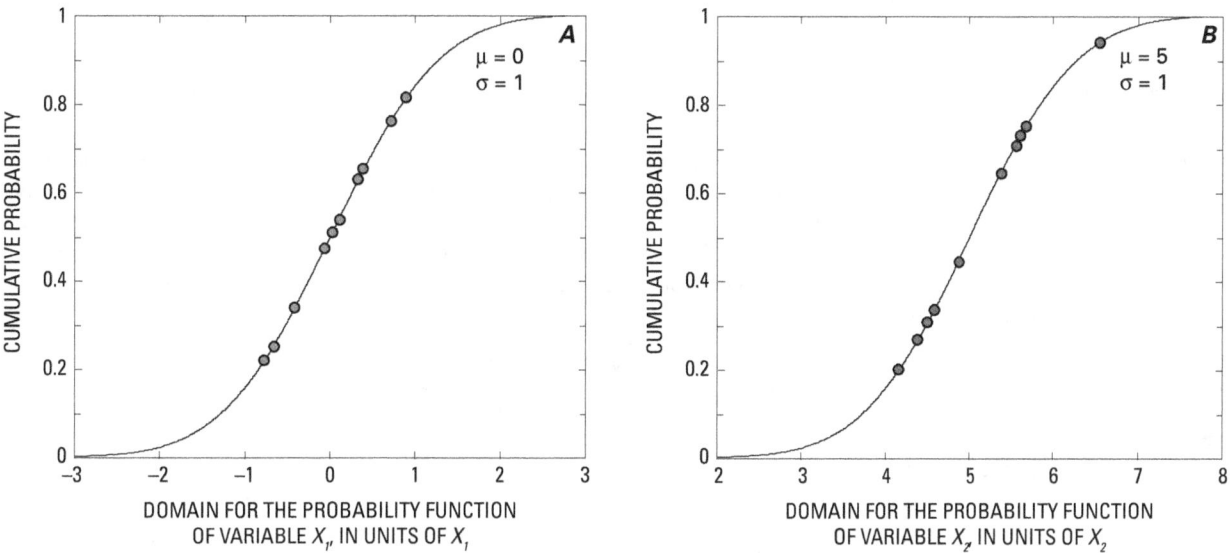

Figure 8.   Randomly selected values using simple random Monte Carlo sampling for (A) X_1 and (B) X_2.

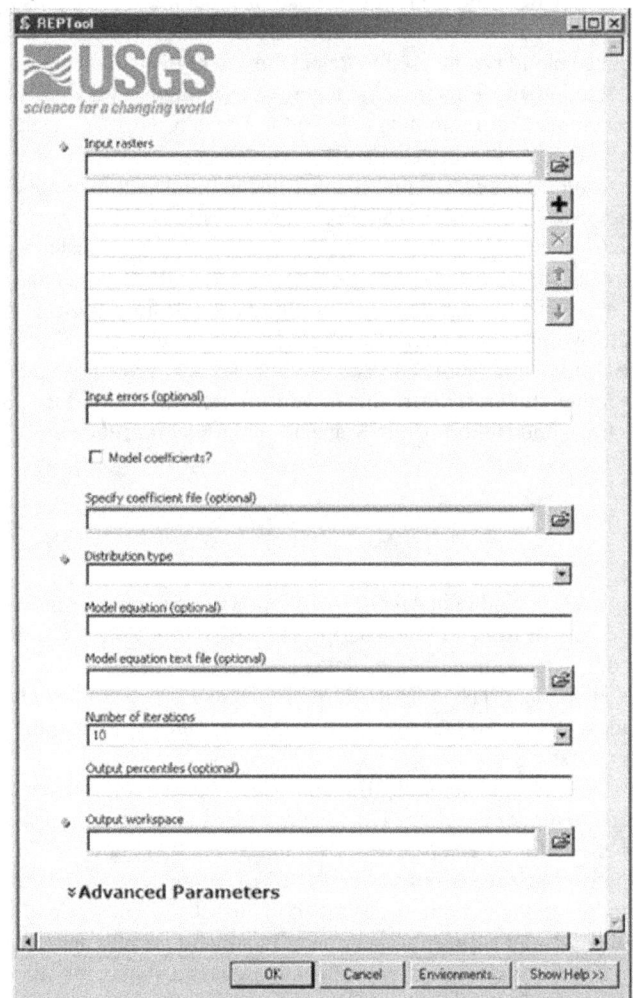

Figure 9.   The dialog window is a graphical user interface
(GUI) for REPTool.

Input Instructions

The following **Input Instructions** describe each user-entry field (fig. 9) that provides input to REPTool. The input user-entry fields are used to specify the geospatial model equation, the inputs and outputs to the model equation, and how error is represented and propagated through the model equation.

The hypothetical model equation and input/output information presented in equation 6 are used as an example throughout many of the subsequent sections of **Input Instructions**. Note—equation 6 is a hypothetical equation presented for illustration purposes and does not represent a real process.

$$\text{Model Equation} = \frac{(c00 * \text{var }00) + (c01 * \text{var }01)}{\text{var }02} \tag{6}$$

where

$\quad$ $c00$ $\quad$ is hypothetical model coefficient 1,

$\quad$ var00 $\quad$ is hypothetical raster (model variable) 1,

$\quad$ $c01$ $\quad$ is hypothetical model coefficient 2,

$\quad$ var01 $\quad$ is hypothetical raster (model variable) 2, and

$\quad$ var02 $\quad$ is hypothetical raster (model variable) 3.

The model variables and coefficients of equation 6 represent the following values:

    $c00$ is 1.158

    var00 is depth to groundwater (dtw) (raster file name: dtw_ex)

    $c01$ is –0.010

    var01 is percentage of irrigated land (irrpct) (raster file name: irrpct_ex), and

    var02 is percentage of clay in the soil (clpct) (raster file name: clpct_ex).

The model variables and coefficients of equation 6 have the following user-specified spatially invariant error:

    $c00$ is 35 percent

    var00 is 10 percent,

    $c00$ is 5 percent,

    var01 is 20 percent, and

    var02 is 30 percent.

Preparing Rasters for REPTool

Some preparation and preprocessing of the input rasters are required prior to use with REPTool. Although other preparation may be required for specific applications of REPTool, the following general preparation guidelines for input rasters ensure proper error propagation through the local functions that are being analyzed with REPTool.

- All rasters used as input for analysis in a particular REPTool execution must have the same spatial properties, resolution, extent, and projection.

- The cells of the input rasters must be aligned because REPTool works on a cell-by-cell basis for local geoprocessing functions. This alignment may be achieved by using one of the raster inputs as a reference and snapping the other raster inputs to this raster. The reference raster used as the snap extent can be set in the "Extent" tab of the "Options" item in the Spatial Analyst drop-down menu in the ArcMap session (fig. 10).

- REPTool allows users to consider spatially variable error that is associated with a raster. Because each cell of a raster contains one attribute value, the spatially variable error values must be specified in a raster separate from the corresponding input raster. Rasters of spatially variable error must also have the same spatial properties and alignment as other input rasters.

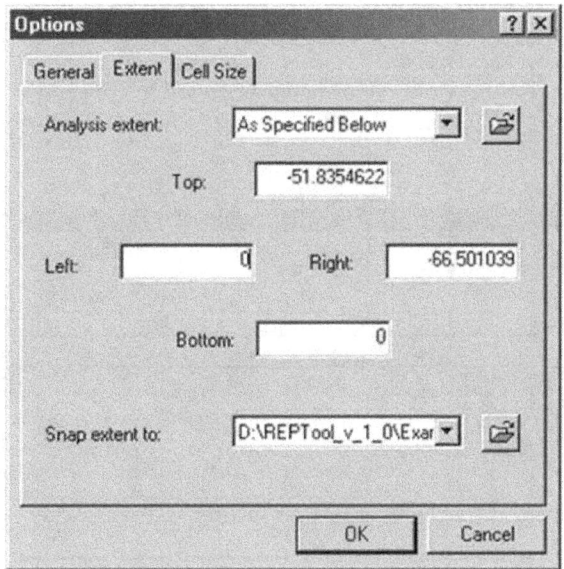

Figure 10.   The "Options" window selected from the Spatial Analyst menu of an ArcMap session.

Note about Computation Runtimes for REPTool:

Before running REPTool, it is important for the user to evaluate the objectives of the geospatial operation or model and the expected computational intensity of REPTool in meeting those objectives. The number of input rasters, the size and resolution of the rasters, the complexity of the model equation, the number of iterations, and the hardware capabilities (that is, one central processing unit [CPU] system compared to multi-CPU systems) of the computer executing REPTool can substantially affect the computation times and the magnitude of CPU and memory usage. As a general guideline, execution times double as the number of iterations doubles (keeping all other input parameters equal). Development testing of REPTool also indicates that each incremental increase in model equation complexity (keeping all other input parameters equal) generally increases execution times by approximately 10 percent. As a specific example of computation times, the model equation in the "**Example Problem**" section ran for approximately 4.3 hours on a computer with hardware capabilities including an Intel Core 2 CPU 6600 @2.4 GHz, 2.39 GHz, and 3.00 GB of RAM. The example model uses four input rasters that each have approximately 7,000 cells at a 500-m resolution, which is approximately 1,750 square kilometers (676 square miles). The model equation is relatively complex and uses 10 iterations for the LHS. Therefore, it is important while evaluating modeling objectives and computational times to consider source error for each input raster and the coarsest resolution that adequately describe the modeled variables to balance model precision with REPTool execution times.

Input Rasters and Errors

All input rasters for REPTool must have the same spatial properties and alignment (see section "**Preparing Rasters for REPTool**"). For convenience and ease of referencing inputs in the model equation, the user is encouraged to select input rasters using the ***Input rasters*** entry field (fig. 11) in the order they appear within the model equation (see section "**Model Equation**").

In the ***Input errors (optional)*** entry field (fig. 11), the user can specify spatially invariant error (that is, error that does not vary spatially) associated with each input raster. The error values must be entered as a percentage (0 to 100) and be separated with commas with no white space in the ***Input errors (optional)*** entry field in the same order as the corresponding input rasters are entered in ***Input rasters*** entry field. The ***Input errors (optional)*** entry field should be left blank if spatially variable errors are assigned to any of the input rasters in the Advanced Parameters of REPTool (see the "**Advanced Parameters**" section of this report for details).

The user-specified error percentage values, *v(x)*, that are entered into the ***Input errors (optional)*** entry field are used in REPTool to define the standard deviation *σ(x)* of *A(x)* (see equation 3). Therefore, *σ(x)* is defined in REPTool as

$$\sigma(x) = \left| \frac{v(x)}{100} * b(x) \right| \qquad (7)$$

where

σ(x)	is the standard deviation of *A(x)*,
v(x)	is the user-defined error (eq. 1), and
b(x)	is the representation of reality (true value) that is expressed by the geospatial data at location *x*.

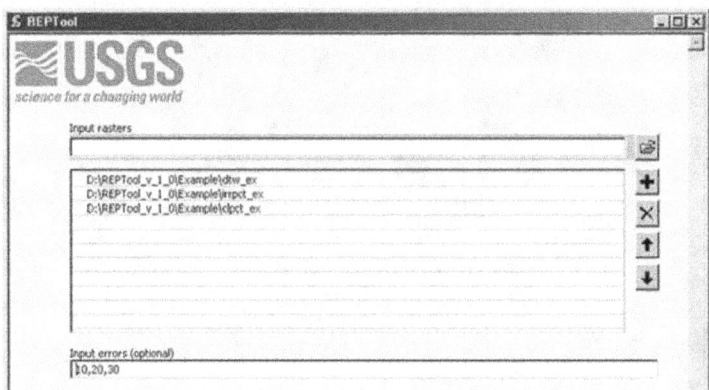

Figure 11.   The ***Input rasters*** and ***Input errors (optional)*** entry fields of REPTool.

It is worth reiterating that REPTool is not designed to explicitly guide or assist the user in identifying, classifying, or quantifying the various types of error that may be associated with the input rasters. Rather, REPTool provides the user a stochastic framework to estimate error in the input rasters, to propagate that error through geospatial processing, and to quantify the effects of errors on the uncertainty of the geospatial-processing results. Therefore, the user has the responsibility of assigning error values, $v(x)$, for input rasters used in REPTool (see "**Theory of Error Propagation and Uncertainty**").

Model Coefficients

After specifying the input rasters and error, the user may specify error information about model coefficients, if applicable to the model equation. Not all model equations may have coefficients. The *Model coefficients?* box (fig. 12) should be checked by the user to attribute error to model coefficients during the REPTool error-propagation calculations. The coefficients and associated error are input to REPTool from a user-generated text (.txt) file that may be selected from the *Specify coefficient file (optional)* entry field (fig. 12).

Similar to the input for error values of rasters (see equation 7 and fig. 11), the error in the coefficient text file must be expressed as error percentage values. The coefficient text file should have the following general format:

<coefficient value>,<error percentage>:
<coefficient value>,<error percentage>:
<coefficient value>,<error percentage>:
<coefficient value>,<error percentage>

Note that each coefficient and error-percentage value is comma separated with no white space, and there is no colon at the end of the last line. The colon is used to separate individual sets of coefficient and error percentage values and is therefore not needed after the last set of values. The example coefficient text file for equation 6 contains:

1.158,35:
0.010,5

Distribution Type

The user must select the shape of the cumulative distributions for the LHS by using the *Distribution type* drop-down field (fig. 13). The current version of REPTool provides Normal, Lognormal, or Uniform distributions for the LHS. The selected distribution shape is applied to the LHS of all uncertain variables and coefficients that are specified in the model equation.

Model Equation

The model equation is specified directly in the *Model equation (optional)* entry field or entered as a text file in the *Model equation text file (optional)* (fig. 14) to facilitate input and entering multiple simulations of a complex model equation. The example model equation (eq. 6) has been entered in the Model equation entry field of figure 14. REPTool will not run if both the *Model equation (optional)* and *Model equation text file (optional)* entry fields are left blank. Therefore, the user must either specify a model equation manually or by using the text file option. Note—the *Model equation text file* option must be used to input the model equation if REPTool is run from the ArcMap command line. The ArcMap command-line interpreter does not accept special characters in model equations, which will cause an error if the model equation is specified using the *Model equation* option. Therefore, it is necessary to input the model equation as a text file when running REPTool from the ArcMap command line.

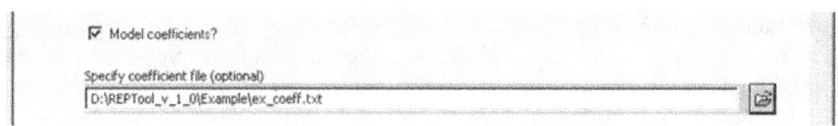

Figure 12. The *Model coefficients?* and *Specify coefficient file (optional)* entry fields of REPTool.

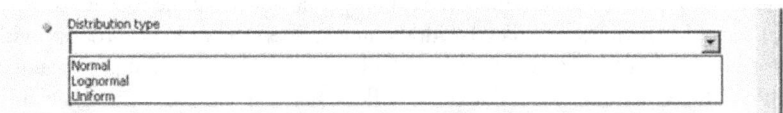

Figure 13.   The ***Distribution type*** drop-down field of REPTool.

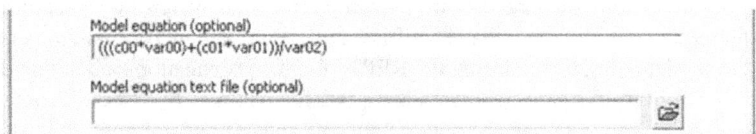

Figure 14.   The ***Model equation (optional)*** and ***Model equation text file (optional)*** entry fields of REPTool.

It is important to note that the math handler for REPTool requires specific syntax and semantics for coefficients, variables, and functions that are expressed in the ***Model equation*** and ***Model equation text file*** entry fields (fig. 14). Therefore, the user is encouraged to spend time becoming familiar with the following section on "**How to Write Model Equations in REPTool,**" which describes in detail the syntax and semantics of writing model equations and naming conventions used in REPTool.

How to Write Model Equations in REPTool

Model equations in REPTool may have zero or more sign characters ('-') followed by an expression (<expr>). An expression may include the following:

an open parenthesis character	('('),
followed by zero or more sign characters	('-'),
followed by a left-hand-side operand	(<operand>),
followed by a binary operator	('-', '+', '*', '/', or '%'),
followed by zero or more sign characters	('-'),
followed by a right-hand-side operand	(<operand>), and
followed by a closing parenthesis character	(')').

The final expression has the general form of:

     <sign(s)> <(> <sign(s)> <operand> <binop><sign(s)> <operand> <)>

Note—illegal characters are treated as white-space characters.

Because REPTool requires at least two raster inputs, the simplest specific expression that may be entered into the equation field is a simple binary operation. Using the syntax required by REPTool, an example of this simple binary operation would be:

$$\left(\text{var}\,00 - \text{var}\,01 \right) \tag{8}$$

where the parentheses around the binary operation are required and the subtraction operator may be any valid operator allowed by REPTool syntax. Another simple example may be the combination of two binary operations:

$$\left(\left(\text{var}\,00 + \text{var}\,01 \right) - \left(\text{var}\,02 + \text{var}\,03 \right) \right) \tag{9}$$

where each binary operation requires a set of parentheses; the two binary operations within the outer parentheses each require a set of parentheses. Because the combination of the two inner binary operations is a binary operation, a set of parentheses must be used around the whole equation.

Users may build more elaborate expressions using substitution of equivalent subexpressions. For example, the following subexpressions may be substituted for any <operand> in an <expr>:

<var>	is a variable,
<literal>	is a literal value,
<func>	is a function, and
<expr>	is another expression

where a function may be:

a function name defined by REPTool syntax	('exp', 'log', and so forth),
followed by an opening parenthesis character	('('),
followed by zero or more sign characters preceding each parameter	('-'),
followed by a one or more parameters separated by commas	(<param>, ...),
followed by a closing parenthesis character	(')').

The final lexeme, which is the lowest-level syntactic or semantic unit of a language, has the general form of:

'function name'<(> [<param> <comma>] ... <)>; and

an example function may have the specific form of:

exp(---10, log(2,5))

Users may build more elaborate functions by using substitution of equivalent subexpressions. For example, the following subexpressions may be substituted for any <param> in a <func>:

<var>	is a variable,
<literal>	is a literal value,
<func>	is another function, and
<expr>	is an expression,

where a variable may be:

range: [var00 to var99],

range: [$c00$ to $c99$], or

range: [a syntax defined constant (e or pi)].

The following example demonstrates five steps to convert a natural equation into REPTool correct form and is slightly more complex than the example provided in equation 6. The example equation for this demonstration is

$$P = \frac{e^{\left[1.158+\left(-0.010*dtw\right)+\left(0.013*irrpct\right)+\left(0.011*clpct\right)\right]}}{1+e^{\left[1.158+\left(-0.010*dtw\right)+\left(0.013*irrpct\right)+\left(0.011*clpct\right)\right]}} \tag{10}$$

where

P	is the predicted probability of the model equation,
e	is the base of natural logarithm,
dtw	is depth to groundwater (dtw),
$irrpct$	is percentage of irrigated land ($irrpct$), and
$clpct$	is percentage of clay in the soil ($clpct$).

Step 1: Begin full qualification.

Full qualification means that the user must specify the order of the operations completely. To fully qualify equation 10, first rename the expression variables to match the syntax and semantics of REPTool following Steps 1 a–c:

a. Replace coefficients with $c00$ to $c99$ depending on the number of coefficients used in the equation and according to the order specified in the coefficient input file (see fig. 12).
b. Replace raster file names with var00 to var99, depending on the number of rasters used in the equation and according to the order specified in the REPTool dialog window (see fig. 11).
c. Replace function names with REPTool names for those functions defined in the equation.

The Steps 1a–c applied to equation 10 result in the following:

$$P = \frac{\exp\left[c00+\left(c01*var00\right)+\left(c02*var01\right)+\left(c03*var02\right)\right]}{1+\exp\left[c00+\left(c01*var00\right)+\left(c02*var01\right)+\left(c03*var02\right)\right]} \tag{11}$$

Step 2: Determine operator precedence.

Determine operator precedence of the example equation (equation 10) according to how the equation is evaluated mathematically. For example, in equation 11 the parenthesis delimiters already surround the coefficient and raster value pairs (that is, $(c01*var00)$). If the parenthesis delimiters were omitted in this equation 11, the operator precedence of a typical calculator would evaluate the multiplications first before any of the additions were made just as if the parenthesis delimiters were present. However, REPTool does NOT apply operator precedence automatically the way a typical calculator does. REPTool requires full qualification of equations and therefore the user must specify the order of operations explicitly and completely.

In equation 11, the set of additions internal to the exp( ) functions have the same precedence so these components do not need further specification. Likewise, the single function in the numerator and the simple expression in the denominator (1 + a function) require no further specification. However, in an example situation where the exp() function contains a set of additions that are divided by another set of additions, the user is required to qualify each of the sets of additions with parenthesis delimiters and then qualify the division with another set of delimiters to produce a REPTool expression.

Step 3: Replace interior components with REPTool syntax and semantic equivalents.

The interior components of the equation must be replaced with REPTool syntax and semantic equivalents. For example, a function in REPTool has a REPTool name identifier, such as 'exp' for the exponential function (equation 11) and a specific set of allowed formats for operations according to the number of parameters the function takes. Specific to equation 11, 'exp' takes only one parameter and raises 'exp' to the power of that parameter. The operational requirements for specifying the parameter are as follows:

a.   In the form: -- exp(param) --
         'exp' names the function and
         -- ( ) --delimiters surround the parameter.
b.   The parameter can be any one of three possibilities: a variable or literal value, another function, or an expression. All three possible versions of a parameter are evaluated before the function is applied. The following are examples of allowable forms for the exp() function:
         exp(25),
         exp(var00),
         exp(log(2.5, 10)), or
         exp(((c00*var00)+(c01*var00))).

Every parameter in a REPTool function must be replaced in the same fashion with inner functions and expressions that conform to the syntax and semantic requirements of REPTool.

In equation 11, the exp() functions contain a literal value and three binary operations connected by addition into an expression that must be translated into REPTool form. Therefore, all binary operations must be surrounded by a parenthesis pair to be evaluated correctly by the math handler used in REPTool. The coefficients and variable (that is, c##*var##) pairs are already formatted. Beginning from left to right, the following consecutive equations (12–14) show how equation 11 is converted to binary operations:

$$P = \frac{\exp\left(\left(c00 + \left(c01 * var\,00\right)\right) + \left(c02 * var\,01\right) + \left(c03 * var\,02\right)\right)}{1 + \exp\left(\left(c00 +\right)\left(c01 * var\,00\right) + \left(c02 * var\,01\right) + \left(c03 * VAR02\right)\right)} \tag{12}$$

$$P = \frac{\exp\left(\left(\left(c00 + \left(c01 * var\,00\right)\right) + \left(c02 * var\,01\right)\right) + \left(c03 * var\,02\right)\right)}{1 + \exp\left(\left(\left(c00 + \left(c01 * var\,00\right)\right) + \left(c02 * var\,01\right)\right) + \left(c03 * var\,02\right)\right)} \tag{13}$$

$$P = \frac{\exp\left(\left(\left(\left(c00 + \left(c01 * var\,00\right)\right) + \left(c02 * var\,01\right)\right) + \left(c03 * var\,02\right)\right)\right)}{1 + \exp\left(\left(\left(\left(c00 + \left(c01 * var\,00\right)\right) + \left(c02 * var\,01\right)\right) + \left(c03 * var\,02\right)\right)\right)} \tag{14}$$

Notice how each left-hand-binary operation from equation 11 becomes a REPTool expression and the left-hand operand of the next binary operation requires conversion.

Step 4: Replace the next operational precedence with REPTool syntax and semantics.

The next level of operational precedence must be replaced with REPTool syntax and semantic elements. For example, equation 12 now has a REPTool correct function divided by an expression that consists of a literal plus, '+', a REPTool correct function. Therefore, the next level of operational precedence for this demonstration example involves translating the expression in the denominator into a REPTool correct expression by placing it inside parenthesis delimiters:

$$P = \frac{\exp\left(\left(\left(\left(c00+\left(c01*\text{var}\,00\right)\right)+\left(c02*\text{var}\,01\right)\right)+\left(c03*\text{var}\,02\right)\right)\right)}{\left(1+\exp\left(\left(\left(\left(c00+\left(c01*\text{var}\,00\right)\right)+\left(c02*\text{var}\,01\right)\right)+\left(c03*\text{var}\,02\right)\right)\right)\right)} \tag{15}$$

Step 5: Continue replacing successive levels of operational precedence.

Each successive level of operational precedence must be replaced until a single REPTool correct expression remains (eq. 14). Therefore, the long bar in equation 13 must be replaced with a division symbol and the entire expression is surrounded by a parenthesis pair:

$$P=(\exp((((c00+(c01*\text{var}00))+(c02*\text{var}01))+(c03*\text{var}02)))/(1+\exp((((c00+(c01*\text{var}00))+(c02*\text{var}01))+(c03*\text{var}02))))) \tag{16}$$

Equation 16 represents the REPTool correct form of equation 10. The user would enter equation 16 into the *Model equation (optional)* or *Model equation text file (optional)* entry fields of REPTool (fig. 14).

Number of Iterations

The number of LHS sample iterations, n (see "**Latin Hypercube Sampling Method**"), allowed in REPTool is 10, 25, 50, or 100 and is selected from a drop-down field (fig. 15). It is important to note that REPTool computation times increase as n increases. An n value of 10 or 25 may be sufficient to adequately describe most CDFs using LHS.

Output Percentiles and Workspace

REPTool requires user specifications about output percentiles and workspace (fig. 16). The output percentiles refer to the percentiles of the distribution surrounding $P(x)$ for each raster cell (see sections "**Application of Quantitative Error Model and Error Propagation in GIS**" and "**Latin Hypercube Sampling Method**" for additional details of the distribution of $P(x)$). REPTool allows the following user-specified output percentile values:
1, 5, 10, 15, 20, 25, 30, 35, 40, 45, 50, 55, 60, 65, 70, 75, 80, 85, 90, 95, and 99.
Additionally, the default output from REPTool provides summary statistics on $P(x)$ for each raster cell and includes minimum (MIN), maximum (MAX), mean (MEAN), median (MEDIAN), and standard deviation (STDEV) values (see "**Description of Output Files**" for additional details on REPTool output).

Multiple output percentile values may be entered in the *Output percentiles (optional)* entry field by the user for any single REPTool run to evaluate statistical properties of $P(x)$ and must be separated by a semicolon with no white spaces (fig. 16). For example, the user may evaluate the minimum (that is, MIN default output), median (that is, enter '50' as the output percentile value), and maximum (that is, MAX default output) values of $P(x)$. Additionally, the user may evaluate various prediction intervals of $P(x)$ uncertainty. For example, the 90-percent prediction interval of $P(x)$ can be evaluated using the 5th and 95th percentile of $P(x)$ by entering '5;95' as the output percentile values (see fig. 16) and then subtracting the 5th percentile values from the 95th percentile values of the REPTool output file by using the Field Calculator application in ArcMap.

The user may specify any output workspace directory in the *Output workspace* entry field of REPTool (fig. 16). Additional details about REPTool output are provided in "**Description of Output Files**."

Figure 15. The *Number of iterations* entry field of REPTool.

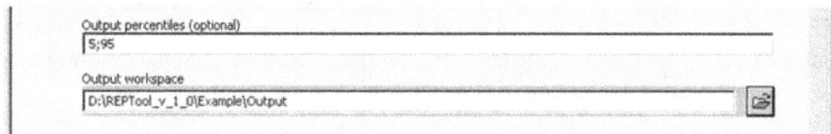

Figure 16.   The *Output percentiles (optional)* and *Output workspace* entry fields of REPTool.

Advanced Parameters

The advanced functions of REPTool are accessed using the optional *Advanced Parameters* drop-down field from the dialog window (fig. 17). The *Advanced Parameters* enable the user to evaluate spatially variable error that may be associated with input variables of the model equation and to evaluate the relative-error contributions from various components of the model equation. The *Advanced Parameters* may not be applicable for some simple operations, particularly if estimates of spatially variable error are not available for input variables. Applications of REPTool to evaluate spatially invariant error will not use the *Advanced Parameters*.

To evaluate the effects of spatially variable error on model output, $P(x)$, using *Advanced Parameters*, the user must specify a separate error raster for each input raster that is specified in the *Input rasters* entry field (fig. 11). The spatial attributes for each error raster are thus estimates of spatially variable error for each corresponding input raster. The error rasters must be entered in *Input spatially variable error rasters (optional)* entry field of REPTool (fig. 17) in the same order as their corresponding input rasters were entered in the *Input rasters* entry field (fig. 11).

The *Advanced Parameters* must be used for model equations that consider spatially variable error for some variables and spatially invariant error for other variables (fig. 17). Because *Advanced Parameters* require a separate error raster for each input raster (that is, each variable in the model), error rasters must be specified for both spatially invariant and spatially variable cases when considering both error types. The spatially invariant error rasters (that is, dtw_err, irrpct_err, and clpct_err in figure 17) will simply have a constant value for all error-attribute values, whereas the spatially variant error rasters (that is, nirrpct_err in figure 17) will have error-attribute values that vary over the spatial domain that are displayed as a raster and input by the user. For example, the error rasters dtw_err, irrpct_err, and clpct_err will each have constant attribute values of 10 percent,

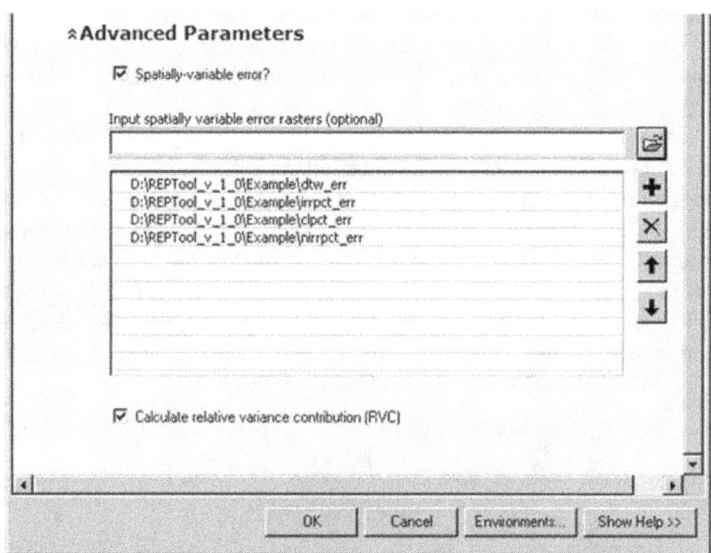

Figure 17.   The *Advanced Parameters* of REPTool includes the *Spatially variable error* check box, *Input spatially variable error rasters (optional)* entry field, and the *Calculate relative variance contribution (RVC)* check box.

20 percent, and 30 percent, respectively, as defined in equation 6 and figure 11. However, the hypothetical variable nirrpct has spatially variable error that must be assigned by the user as an error raster (nirrpct_err in figure 17) for input to REPTool. The user is encouraged to read the "**Raster Processing, Error, and Uncertainty**" section of this manual in considering the effects of spatially variable error on model output and assigning those error values to REPTool *Advanced Parameters*.

The relative variance contribution (RVC) analyses are run by checking the *Calculate relative variance contribution* box in the *Advanced Parameters* of REPTool (fig. 17). The RVC analysis is done on each cell of *P(x)* and determines the RVC of error from the model coefficients and from model variables (input rasters). The RVC calculations in REPTool are based on a method originally presented by van Horssen and others (2002) to evaluate spatial interpolation during ordinary block kriging.

Using the RVC approach (van Horssen and others, 2002), the total prediction variance of *P(x)*, expressed as $\sigma^2[P(x)]$, is equal to the sum of the variance as a result of error in the model coefficients, $\sigma_c^2[P(x)]$, and variance as a result of error in model variables, $\sigma_v^2[P(x)]$. The general decomposition of the total prediction variance is as follows:

$$\sigma^2\left[P(x)\right] = \sigma_v^2\left[P(x)\right] + \sigma_c^2\left[P(x)\right] + 2*Cov\left\{\sigma_v^2\left[P(x)\right], \sigma_c^2\left[P(x)\right]\right\} \tag{17}$$

where

$\sigma^2[P(x)]$ is the total prediction variance of *P(x)*,

$\sigma_v^2[P(x)]$ is the variance of *P(x)* as a result of error in the model variables,

$\sigma_c^2[P(x)]$ is the variance of *P(x)* as a result of coefficients in the model variables,

and

$Cov\{\sigma_v^2[P(x)], \sigma_c^2[P(x)]\}$ is the covariance of each pair of $\sigma_v^2(P(x)$ and $\sigma_c^2(P(x))$ that are formed from the components of the sum.

The covariance of $\sigma_v^2[P(x)]$ and $\sigma_c^2[P(x)]$ is assumed in REPTool to be negligible because of independence between components in the model equation. Therefore, the relative variance contribution due to the model variables (RVC_v) is calculated as follows:

$$RVC_v = \frac{\sigma_v^2\left[P(x)\right]}{\sigma^2\left[P(x)\right]} * 100 \tag{18}$$

The relative variance contribution due to model coefficients (RVC_c) is calculated as follows:

$$RVC_c = \frac{\sigma_c^2\left[P(x)\right]}{\sigma^2\left[P(x)\right]} * 100 \tag{19}$$

If the error from the model variables and coefficients contribute equally to the uncertainty of *P(x)*, the RVC_v will equal the RVC_c. However, values of RVC_c greater than RVC_v indicate locations of the spatial domain, *D*, where uncertainty due to model coefficients dominates the total uncertainty of *P(x)*. This result would indicate the need for improved characterization of the model in those locations to reduce the total prediction uncertainty of *P(x)*. For example, Gurdak and others (2007) indicated that RVC_c may be improved by installing additional monitoring wells to better calibrate coefficients for a logistic regression-based groundwater vulnerability model. RVC_c values less than the RVC_v values indicate locations of *D* where *P(x)* uncertainty due to error in model variables dominates the total prediction uncertainty. This result would indicate the need to reduce error in input rasters in those locations to reduce the total prediction uncertainty of *P(x)*.

If the *Calculate relative variance contribution* box is checked in the *Advanced Parameters*, the output from REPTool includes $\sigma^2[P(x)]$, $\sigma_v^2[P(x)]$, $\sigma_c^2[P(x)]$, RVC_v, and RVC_c. As stated above, $\sigma^2[P(x)]$ is calculated as the variance of the model output distribution, *P(x)*, after LHS at each cell in the output raster. During the REPTool calculations of $\sigma^2[P(x)]$, the model variables, $A_i(x)$, and coefficients, $b_i(x)$, (see equation 5) are treated as distributions in the LHS as specified by user input (see equation 7). However, during the REPTool calculations of $\sigma_v^2[P(x)]$, the model variables are treated as distributions, $A_i(x)$, in the LHS, but the $b_i(x)$ are treated as the model-coefficient values specified in the model-coefficient input file. Therefore, $\sigma_v^2[P(x)]$ represents the variance in *P(x)* that is the result of error introduced only by the model variables. Conversely, during the REPTool calculations of $\sigma_c^2[P(x)]$, the model coefficients, $b_i(x)$, are treated as distributions in the LHS, but the model variables are treated as the attribute values, $b(x)$ (see equation 1), in each raster. Therefore, $\sigma_c^2[P(x)]$ represents the variance in *P(x)* that is the result of error introduced only by the model coefficients. Assuming independence between the model variables and coefficients, the $\sigma^2[P(x)]$ should equal $\sigma_v^2[P(x)]$ plus $\sigma_c^2[P(x)]$ (equation 17). However, discrepancies in this logic have been shown to occur because of the nature of random pairing in LHS (Iman and Conover, 1982) (see the "**Latin Hypercube Sampling Methods**" section of this manual) and (or) because all of the variance and covariance components of the model equation may not be accounted for, and the assumptions of independence between model variables and coefficients may not be appropriate (Thorsen and others, 2001). In spite of this, Thorsen and others (2001) and Gurdak and others (2007) suggest that the RVC approach may still provide an indication of the relative importance of the error introduced from the sum of model variables and the error introduced from the sum of model coefficients on the overall model-prediction uncertainty. Additional details about the RVC in the REPTool output are presented in "**Description of Output Files.**"

Description of Output Files

The output from REPTool is stored in the *REPData* directory that is created by REPTool under the user-specified output directory entered in the **Output workspace** entry field (fig. 16). Each run of REPTool will create a subdirectory under the *REPData* directory that is named with a *timestamp* generated at the start of the REPTool execution.

REPTool also creates a log-file directory in *C:/temp/logs* and files for logging errors, *log_003_tool_<timestamp>.txt*, where *timestamp* is the actual time stamp of the execution. If no errors are generated during a REPTool execution, the log file will only contain the execution start and the end time. The log files may be useful for reporting results and determining errors.

Under the <output workspace>\REPData\timestamp subdirectory, REPTool creates three subdirectories for each execution: *temp*, *data*, and *results*. Although the *temp* and *data* subdirectories contain intermediate files that may be useful for some applications, the *results* subdirectory contains the final results from the model equation and will be important for all REPTool runs. A brief description of *temp* and *data* subdirectories is presented next, followed by a detailed description of the *results* subdirectory.

Under the *temp* subdirectory, REPTool generates a *tempgrd.shp* point shapefile and set of associated files and folders that ArcMap generates automatically during raster to point shapefile conversions. The *temp* subdirectory also contains four text files, *rawsource.txt*, *synlex.txt*, *semlex.txt*, and *execf.txt*, that are generated during Map Algebra processing. The *rawsource.txt* file contains the REPTool-correct version of the model equation that was used by REPTool, which may be useful in troubleshooting REPTool execution errors. The other files in the *temp* subdirectory are not needed for normal operation of REPTool and may be ignored for most REPTool applications.

Under the *data* subdirectory, REPTool generates a subdirectory for each input raster variable name and associated spatially variable error raster files. For example, a model equation that has three input variables and three spatially variable error files will generate *var00, err00, var01, err01, var02,* and *err02* subdirectories. Each of these subdirectories will contain a *shp* subdirectory that stores a point shapefile and associated ArcMap files with the same name as the *var* or *err* subdirectory. The point shapefiles are created during the previously described raster-to-point conversion by REPTool and contain the values for each raster cell. The *var* subdirectories (that is, *var00, var01, var02,* and so forth) will also contain two text files that are suffixed with the variable name before the .txt extension and prefixed by *dist_* or *std_dev_*, respectively (that is, *dist_var00.txt* and *std_dev_var00.txt* under a *var00* subdirectory). These files store distribution and standard-deviation calculation results and serve as input files for execution of the model equation. The information stored in the *data* subdirectories may be ignored by users for most REPTool applications. However, some users may find useful information in the distribution and standard deviation intermediate files.

Under the *results* subdirectory, REPTool generates the results of the model equation, the output percentile calculations, and the RVC calculations, which are all stored in the *results.shp* point shapefile. The points in *results.shp* represent the center of each cell in the input raster data. The attribute list for each point in *results.shp* includes default outputs and user-specified outputs that are based on the **Input Instructions** used to run REPTool. The default output includes FID, Shape, POINTID, GRID_CODE, and summary statistics about *P(x)* for each raster cell in the model output (fig. 18). The summary statistics for *P(x)* include minimum (MIN), maximum (MAX), mean (MEAN), median (MEDIAN), and standard deviation (STDEV) values (fig. 18).

The *results.shp* attribute fields that appear to the right of the STDEV field are dependent on the user-specified input instructions to REPTool. User-requested output percentiles (see "**Output Percentiles and Workspace**") will be named "percX" or "percXX" where "X" or "XX" is the value of the output percentile requested (fig. 18). Additional attribute fields in *results.shp* include the RVC fields (VARtot, VARv, VARc, RVCv, and RVCc) (fig. 18). The VARtot field represents $\sigma^2[P(x)]$; VARv field represents $\sigma_v^2[P(x)]$; VARc field represents $\sigma_c^2[P(x)]$; RVCv field represents RVC_v; and the RVCc field represents RVC_c (see equations 17–19). The RVC fields are explained in the section "**Advanced Parameters.**"

The user may convert any of the *results.shp* point attributes into a raster using the Features to Raster conversion in the Spatial Analyst extension in ArcGIS (fig. 19). Figure 19 illustrates the example conversion of the MEDIAN point attributes to a MEDIAN raster. The MEDIAN raster is shown in figure 20.

Example Problem

In order to demonstrate the applicability and capabilities of REPTool, a slightly more complex simulation is presented here than was used in the "**Input Instructions**" section. The example problem (equation 20) is based on a logistic regression-based groundwater vulnerability model that was originally presented by Gurdak and Qi (2006) and Gurdak and others (2007).

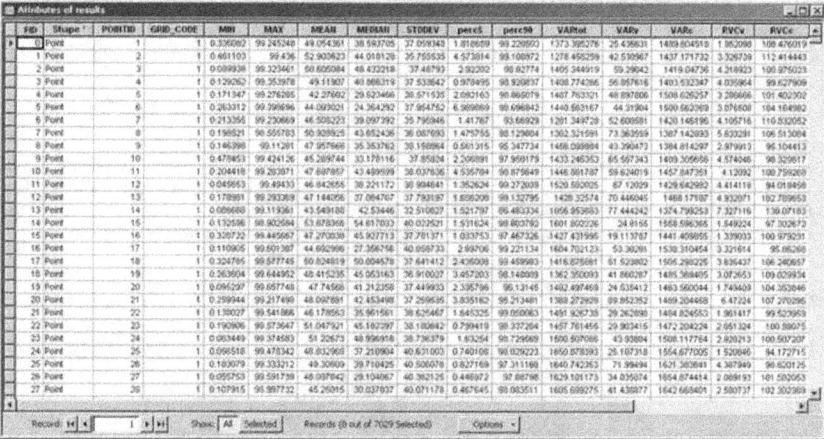

Figure 18. An example output from REPTool that illustrates default values (FID, Shape, POINTID, GRID_CODE) and summary statistics [minimum (MIN), maximum (MAX), MEAN, MEDIAN, and standard deviation (STDDEV)]. Note — for this example the user did not check the ***Model coefficients?*** box, specified ***Output percentile (optional)*** entry values of 5 and 90, and checked the ***Calculate relative variance contribution (RVC)*** box in the ***Advanced Parameters*** of REPTool.

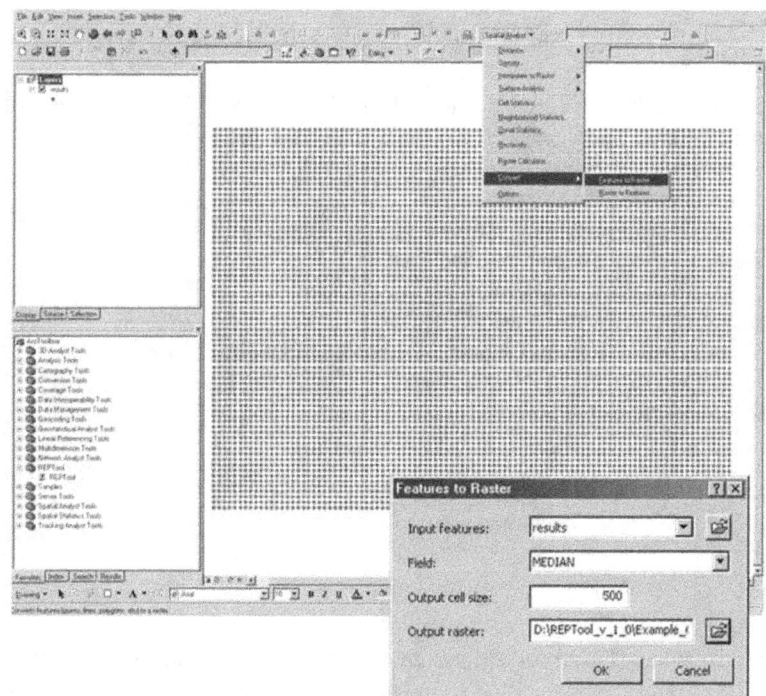

Figure 19. An example of the Features to Raster conversion in the Spatial Analyst extension in ArcGIS using the MEDIAN attribute value from *results. shp.*

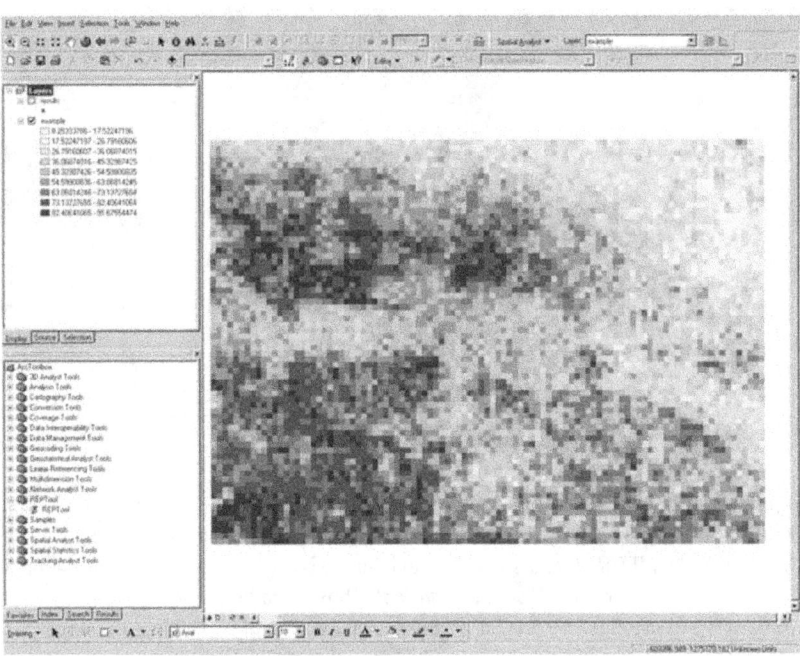

Figure 20.   The MEDIAN raster shown in an ArcMap application after conversion from *results.shp*.

$$P = \left(\frac{e^{\left[1.158+(-0.010*dtw)+(0.013*nirrpct)+(0.011*irrpct)+(-0.019*clpct)\right]}}{1+e^{\left[1.158+(-0.010*dtw)+(0.013*nirrpct)+(0.011*irrpct)+(-0.019*clpct)\right]}} \right) * 100 \tag{20}$$

where

P is the predicted probability of groundwater vulnerability to elevated nitrate concentrations greater than or equal to 4 milligrams per liter (as N), in units of percentage,

e is the base of natural logarithm,

dtw   is depth to groundwater (var00), in meters,

nirrpct   is percentage of nonirrigated land (var01),

irrpct   is percentage of irrigated agricultural land (var02),

and

clpct   is percentage of clay in the soil (var03).

The REPTool-correct version of equation 20 that is input as the model equation text file is:
((exp(((((c00 + (c01*var00))) + (c02*var01)) + (c03*var02)) + (c04*var03))) / (1 + exp(((((c00 + (c01*var00)) + (c02*var01)) + (c03*var02)) + (c04*var03)))))*100)

The associated model-coefficients file is as follows:

    1.158,219:
    –0.010,92:
    0.013,139:
    0.011,186:
    –0.019,179

A normal distribution is used to estimate the error distributions, and the minimum number of iterations (*n*=10) is used for the LHS. Output percentiles of 25, 50, and 75 and the RVC analysis are also specified in the example problem. The input instructions for the example problem are shown in the REPTool dialog window (fig. 21).

   The results of the example problem indicate spatial variability in the predicted probability of groundwater vulnerability to elevated nitrate and in the uncertainty surrounding the vulnerability calculation (fig. 22). The 25th and 75th percentiles of the output distribution represent a 50-percent prediction uncertainty surrounding the mean (50th percentile) model result (fig. 22).

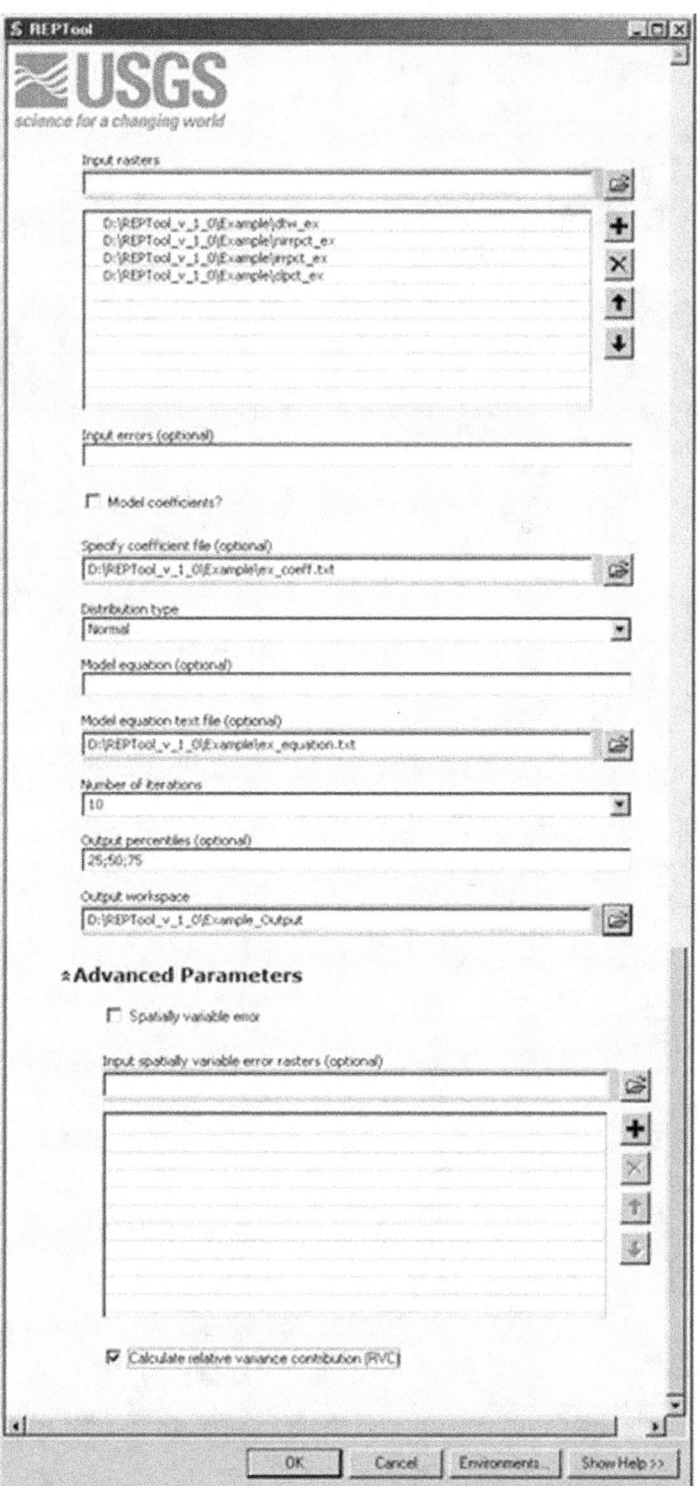

Figure 21. Input instructions in the REPTool dialog window used for the example problem.

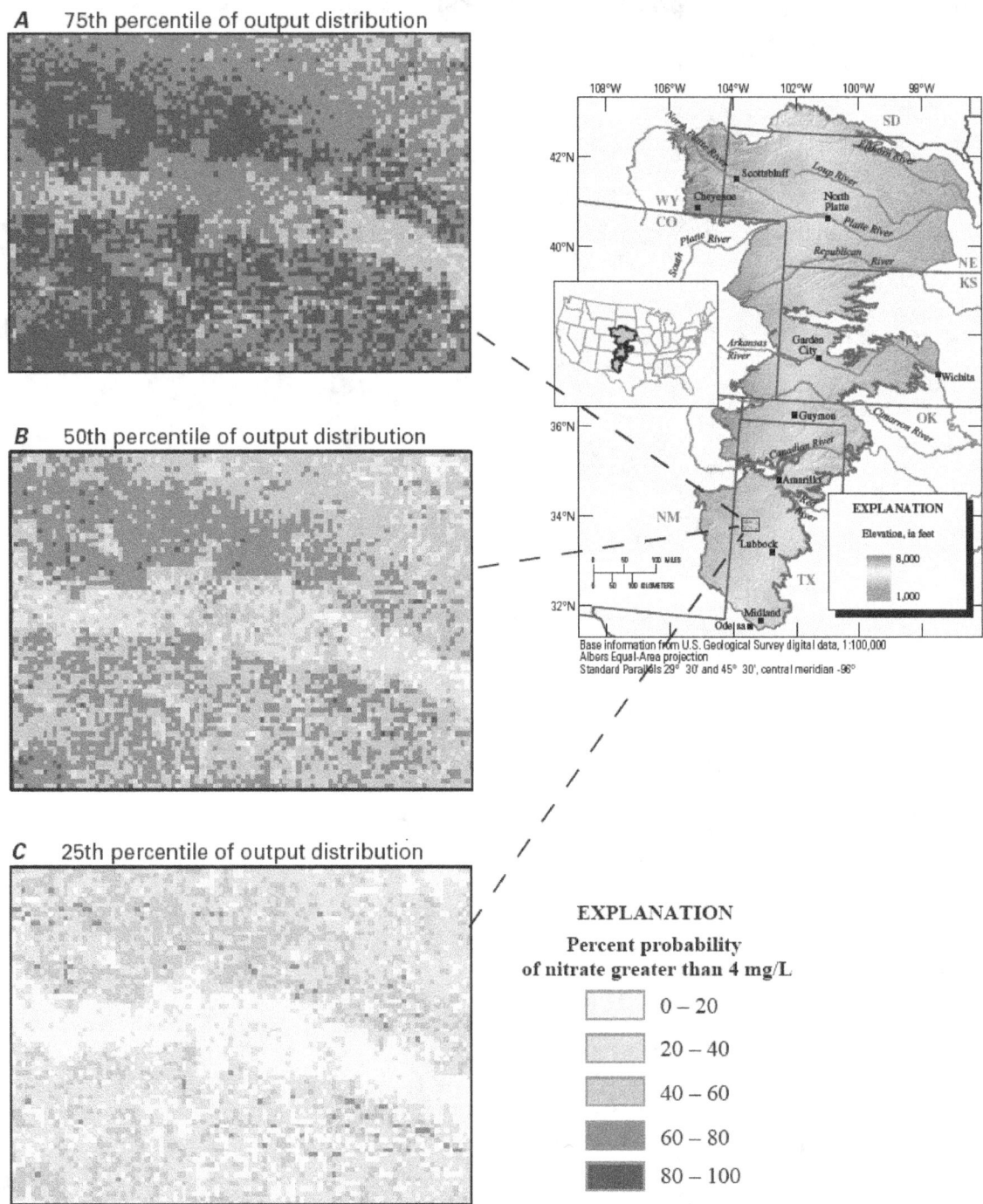

Figure 22. The spatial distribution of the probability of detecting nitrate (as N) greater than 4 milligrams per liter (mg/L) in recently recharged groundwater from the output distribution for (A) 75th percentile, (B) 50th percentile, and (C) 25th percentile.

A prediction interval may be a more meaningful visualization of the uncertainties surrounding a REPTool model result than individual percentiles from the output distribution (fig. 22). A prediction interval may be calculated as the difference between any two output percentile values using the Field Calculator application in ArcMap or the Raster Calculator in Spatial Analyst. For example, the 50-percent prediction interval of uncertainty surrounding the mean (50th percentile) model result is calculated as the difference between the 75th and 25th percentiles (fig. 23). Figure 23A, which is the same map shown in figure 22B, indicates a horizontal band of relatively low probability of nitrate greater than 4 mg/L and relatively higher probability north and south of the horizontal band. However, the 50-percent prediction interval (fig. 23B) is relatively larger in the horizontal band than areas to the north and indicates greater uncertainties in the model results. The information in figure 23A may be useful to groundwater managers interested in knowing where to allocate resources for groundwater protection. The information in figure 23B is equally valuable to resource managers because it provides information about the confidence associated with model results (fig. 23A).

The RVC analysis (fig. 24) indicates that the uncertainty in the model result is dominated by error introduced from the model coefficients because the RVC_c values (fig. 24B) are substantially greater than the RVC_v values (fig. 24A). The type of information provided by the RVC analysis in this example problem could be used by resource managers and scientists to improve the prediction confidence of future versions of the groundwater vulnerability model. The RVC_c could be reduced in the example problem by installing more monitoring wells to reduce the error attributed to the model coefficients (Gurdak and others, 2007). Assuming actions were taken that reduced the error contributions from the model coefficients by a factor of 10, the RVC from the new REPTool output indicate a substantial reduction in RVC_c and corresponding increase in RVC_v (fig. 25).

Most important, the hypothetical reduction in model-coefficient error by a factor of 10 substantially improves the 50-percent prediction interval surrounding the median model results (fig. 26). Prior to the reduction in model-coefficient error, the 50-percent prediction interval was in the range of 20–60 percent (fig. 23B). After the reduction in model-coefficient error, the 50-percent prediction interval is less than 20 percent over most of the spatial domain (fig. 26B). Therefore, the confidence in the example groundwater vulnerability model is substantially improved by reducing the error associated with the model coefficients.

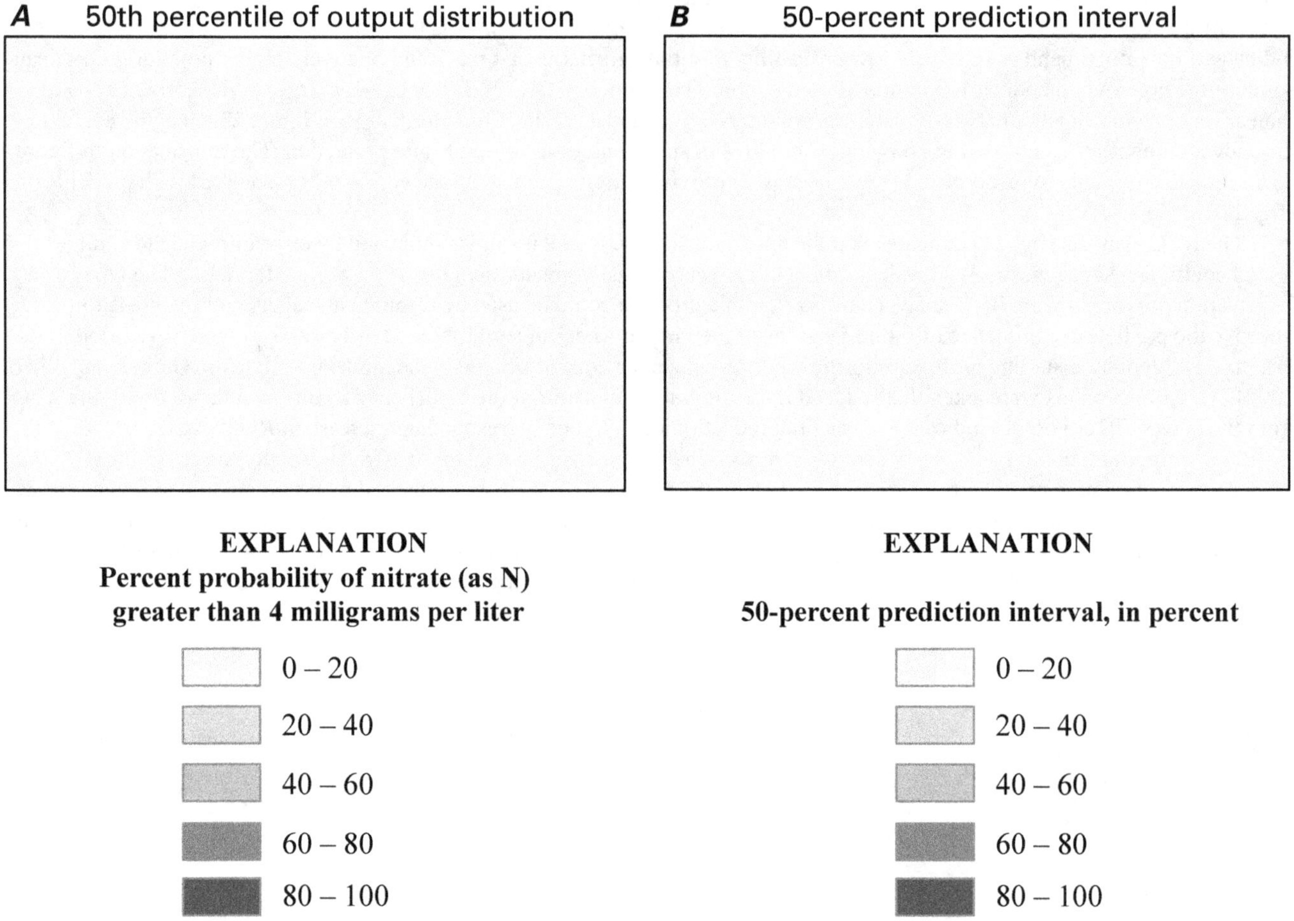

A   50th percentile of output distribution

B   50-percent prediction interval

EXPLANATION
**Percent probability of nitrate (as N)
greater than 4 milligrams per liter**

- [ ] 0 – 20
- [ ] 20 – 40
- [ ] 40 – 60
- [ ] 60 – 80
- [ ] 80 – 100

EXPLANATION

50-percent prediction interval, in percent

- [ ] 0 – 20
- [ ] 20 – 40
- [ ] 40 – 60
- [ ] 60 – 80
- [ ] 80 – 100

Figure 23.   The (*A*) median (50th percentile) model output is surrounded by the (*B*) 50-percent prediction interval, which is the difference between the 75th and 25th percentiles from the model output probability distribution.

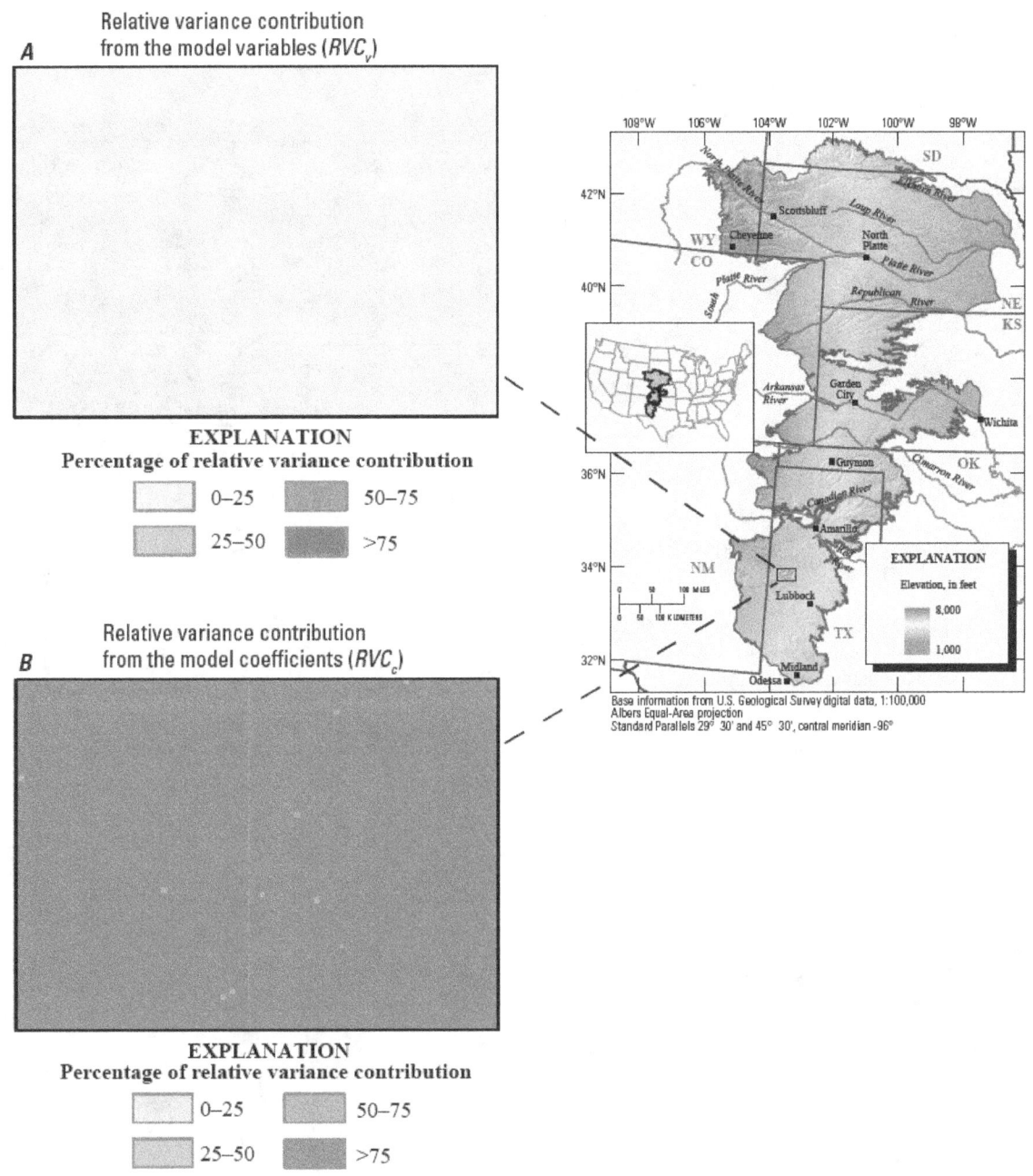

Figure 24. The spatial distribution of the relative variance contributions from the (A) model variables (RVC_v) and (B) model coefficients (RVC_c) based on error introduced from original model coefficients.

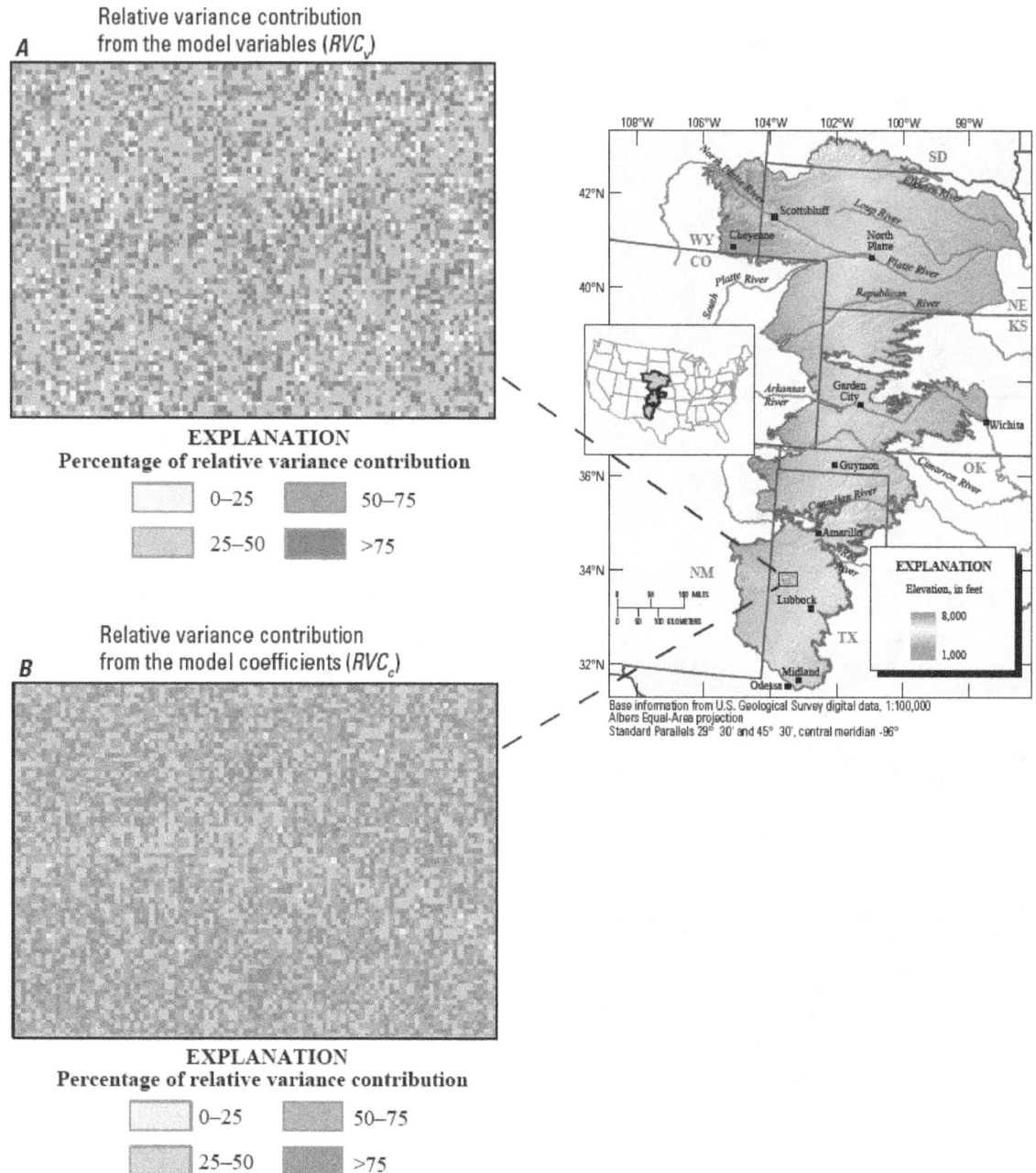

Figure 25.   The spatial distribution of the relative variance contributions from the (A) model variables (RVC_v) and (B) model coefficients (RVC_c) based on error introduced from model coefficients reduced by a factor of 10.

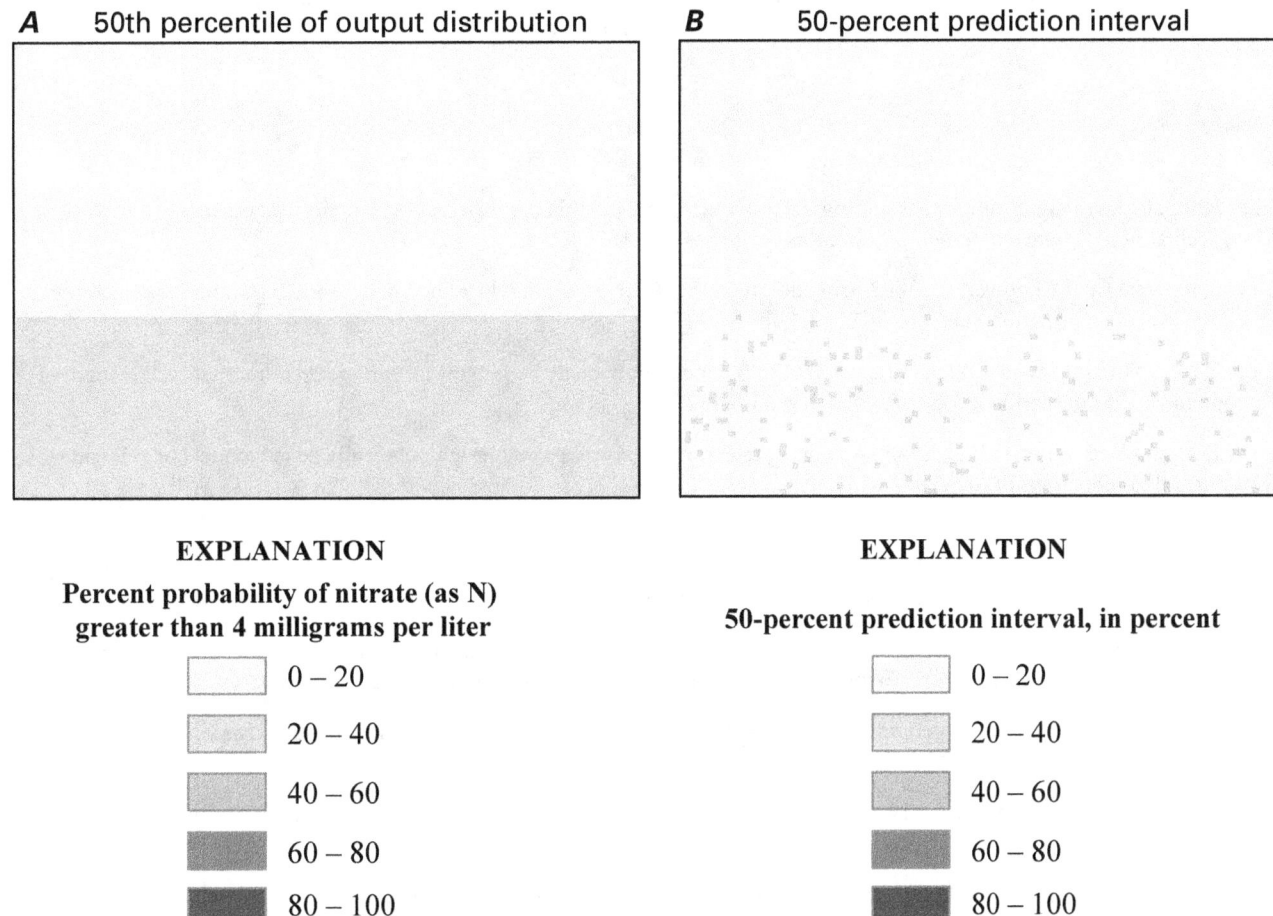

EXPLANATION

**Percent probability of nitrate (as N)
greater than 4 milligrams per liter**

0 – 20

20 – 40

40 – 60

60 – 80

80 – 100

EXPLANATION

50-percent prediction interval, in percent

0 – 20

20 – 40

40 – 60

60 – 80

80 – 100

Figure 26.   The (*A*) median (50th percentile) model output is surrounded by the (*B*) 50-percent prediction interval after a hypothetical reduction in model-coefficient error by a factor of 10.

Acknowledgments

The development of REPTool and this report was supported by the USGS Center of Excellence for Geospatial Information Science (CEGIS). The authors thank Dr. Quan Qi for statistical advice and Virginia L. McGuire, David K. Mueller, Suzanne S. Paschke, Evan Thoms, and Roland Viger (USGS) for helpful reviews of this report and beta testing of REPTool.

References Cited

Acklam, P.J., 2004, An algorithm for computing the inverse normal cumulative distribution function: Accessed on the World Wide Web at *http://home.online.no/~pjacklam/notes/invnorm/index.html on July 14*, 2008.

Bachman, Andreas, and Allgöwer, B., 2002, Uncertainty propagation in wildland fire behavior modeling: International Journal of Geographical Information Science, v. 16, no. 2, p. 115–127.

Bishop, T.F.A., Minasny, B., and McBratney, A.B., 2006, Uncertainty analysis for soil-terrain models: International Journal of Geographical Information Science, v. 20, no. 2, p. 117–134.

Brown, J.D., and Heuvelink, G.B.M, 2005, Assessing uncertainty propagation through physically based models of soil water flow and solute transport, *in* Anderson, M.G., ed., Encyclopedia of Hydrologic Sciences: Chichester, UK, Wiley, p. 1181–1195.

Brown, J.D., and Heuvelink, G.B.M, 2007, The data uncertainty engine (DUE)—A software tool for assessing and simulating uncertain environmental variables: Computers and Geosciences, v. 33, no. 2, p. 172–190.

Buckey, D.J., 2008, The GIS Primer, An introduction to geographic information systems, Vector and raster—Advantages and disadvantages: accessed on the World Wide Web on June 29, 2008, at *http://bgis.sanbi.org/gis-primer/page_19.htm*.

Burrough, P.A., 1986, Principles of geographical information systems for land resources assessment: Oxford, UK, Clarendon Press, 194 p.

Burrough, P.A., and McDonnell, R.A., 1998, Principles of geographical information systems: New York, Oxford University Press, 333 p.

Canters, Frank, De Genst, W., and Dufourmont, H., 2002, Assessing effects of input uncertainty in structural landscape classification: International Journal of Geographical Information Systems, v. 16, no. 2, p. 129–149.

Emmi, P.C., and Horton, C.A., 1995, A Monte Carlo simulation of error propagation in a GIS-based assessment of seismic risk: International Journal of Geographical Information Systems, v. 9, no. 4, p. 447–461.

Environmental System Research Institute (ESRI), 2006, GIS Dictionary: Redlands, Calif., ESRI Support Center, accessed on the World Wide Web at *http://support.esri.com/index.cfm?fa=knowledgebase.gisDictionary.gateway*, on June 23, 2008.

Finke, P.A., Wösten, J.H.M., and Jansen, M.J.W., 1996, Effects of uncertainty in major input variables on simulated functional soil behavior: Hydrological Processes, v. 10, no. 5, p. 661–669.

Fisher, P.F., 1991, Modeling soil map-unit inclusions by Monte Carlo simulation: International Journal of Geographical Information Systems, v. 5, no. 2, p. 193–208.

Foody, G.M., and Atkinson, P.M., eds., 2002, Uncertainty in remote sensing and GIS: Chichester, UK, Wiley, 307 p.

Goodchild, M.F., Sun, G., and Yang, S., 1992, Development and test of an error model for categorical data: International Journal of Geographic Information Science, v. 6, no. 2, p. 87–104.

Gottsegen, Jonathan, Montello, D., and Goodchild, M., 1999, A comprehensive model of uncertainty in spatial data, *in* Lowell, K., and Jaton, A., eds., Spatial accuracy assessment — Land information uncertainty in natural resources: Chelsea, Mich., Ann Arbor Press, 455 p.

Gurdak, J.J., 2008, Ground-water vulnerability: Nonpoint-source contamination, climate variability, and the High Plains aquifer: Saarbrucken, Germany, VDM Verlag Publishing, ISBN: 978-3-639-09427-5, 223 p.

Gurdak, J.J., McCray, J.E., Thyne, G.D., and Qi, S.L., 2007, Latin hypercube approach to estimate uncertainty in ground water vulnerability: Ground Water, v. 45, no. 3, p. 348–361.

Gurdak, J.J., and Qi, S.L., 2006, Vulnerability of recently recharged ground water in the High Plains aquifer to nitrate contamination: U.S. Geological Survey Scientific Investigations Report 2006–5050, 39 p.

Helton, J.C., and Davis, F.J., 2003, Latin hypercube sampling and the propagation of uncertainty in analysis of complex systems: Reliability Engineering and System Safety, v. 81, no. 1, p. 23–69.

Heuvelink, G.B.M., 1996, Identification of field attribute error under different models of spatial variation: International Journal of Geographical Information Science, v. 10, no. 8, p. 921–936.

Heuvelink, G.B.M., 1998, Error propagation in environmental modeling with GIS: London, UK, Taylor and Francis, Ltd., 127 p.

Heuvelink, G.B.M., 1999, Propagation of error in spatial modeling with GIS, in Longley, P.A., Goodchild, M.F., Maguire, D.J., and Rhind, D.W., eds., Geographical information systems, volume 1—Principles and technical issues: New York, John Wiley and Sons, p. 207–217.

Heuvelink, G.B.M., Brown, J.D., and Van Loon, E.E., 2007, A probabilistic framework for representing and simulating uncertain environmental variables: International Journal of Geographical Information Science, v. 21, no. 5, p. 497–513.

Heuvelink, G.B.M., and Burrough, P.A., 1993, Error propagation in cartographic modeling using Boolean logic and continuous classification: International Journal of Geographical Information Systems, v. 7, no. 3, p. 213–246.Heuvelink, G.B.M., Burrough, P.A., and Stein, A., 1989, Propagation of errors in spatial modeling with GIS: International Journal of Geographical Information Systems, v. 3, no. 4, p. 303–322.

Heuvelink, G.B.M., Burrough, P.A., and Stein, A., 1989, Propagation of errors in spatial modeling with GIS: International Journal of Geographical Information Systems, v. 3, no. 4, p. 303–322.

Hunsaker, C.T., Goodchild, M.F., Friedl, M.A., and Case, T.J., 2001, Spatial uncertainty in ecology, Implications for remote sensing and GIS applications: New York, Springer, 402 p.

Iman, R.L., and Conover, W.J., 1982, A distribution-free approach to inducing rank correlation among input variables: Communications in Statistics — Simulation and Computation, v. 11, no. 3, p. 311–334.

Iman, R.L., and Helton, J.C., 1988, An investigation of uncertainty and sensitivity analysis techniques for computer models: Risk Analysis, v. 8, no. 1, p. 71–90.

Krivoruchko, Konstantin, and Gotway, C.A., 2005, Assessing the uncertainty resulting from geoprocessing operations, in Maguire, D.J., Batty, M., and Goodchild, M.F., eds., GIS, spatial analysis, and modeling: Redlands, Calif., ESRI Press, p. 67–92.

Longley, P.A., Goodchild, M.F., Maguire, D.J., and Rhind, D.W., 1999, Data quality introduction, in Longley, P.A., Goodchild, M.F., Maguire, D.J., and Rhind, D.W., eds., Geographical information systems, volume 1—Principles and technical issues: New York, John Wiley and Sons, p. 175–176.

Manache, Gemma, and Melching, C.S., 2004, Sensitivity analysis of a water-quality model using Latin Hypercube Sampling: Journal of Water Resources Planning and Management, v. 130, no. 3, p. 232–242.

McKay, M.D., Beckman, R.J., and Conover, W.J., 1979, A comparison of three methods for selecting values of input variables in the analysis of output from a computer code: Technometrics, v. 21, p. 239–245.

Mowrer, H.T., and Congalton, R.G., eds., 2000, Quantifying spatial uncertainty in natural resources—Theory and applications for GIS and remote sensing: Chelsea, Mich., Ann Arbor Press, 244 p.

Phillips, D.L., and Marks, D.G., 1996, Spatial uncertainty analysis—Propagation of interpolation errors in spatially distributed models: Ecological Modeling, v. 91, no. 1–3, p. 213–229.

Qi, S.L., and Gurdak, J.J., 2006, Percentage of probability of nonpoint source nitrate contamination of recently recharged ground water in the High Plains aquifer: U.S. Geological Survey Data Series, available at *http://water.usgs.gov/lookup/getspatial?ds192_hp_npctprob.*

Sanchez, M.A., and Blower, S.M., 1997, Uncertainty and sensitivity analysis of the basic reproductive rate: American Journal of Epidemiology, v. 145, no. 12, p. 1127–1137.

Sklar, F.H., and Hunsaker, C.T., 2001, The use and uncertainties of spatial data for landscape models—An overview with examples from the Florida Everglades, *in* Hunsaker, C.T., Goodchild, M.F., Friedl, M.A., and Case, T.J., eds., Spatial uncertainty in ecology—Implications for remote sensing and GIS applications: New York, Springer, 402 p.

Swartzman, G.L., and Kaluzny, S.P., 1987, Ecological simulation primer: New York, MacMillan Publishing Company, 370 p.

Swiler, L.P., and Wyss, G.D., 2004, A user's guide to Sandia's Latin Hypercube Sampling Software — LHS UNIX library/stand-alone version: Sandia National Laboratory SAND2004–2439, 82 p.

Thorsen, M., Refsgaard, J.C., Hansen, S., Pebesma, E., Jensen, J.B., and Kleeschulte, S., 2001, Assessment of uncertainty in simulation of nitrate leaching to aquifers at catchment scale: Journal of Hydrology, v. 242, no. 3–4, p. 210–227.

Tomlin, C.D., 1990, Geographic information systems and cartographic modeling: Englewood Cliffs, N.J., Prentice-Hall, 572 p.

U.S. Geological Survey, 1997, Standards for digital elevation models, part 1 general: Reston, Va., U.S. Department of the Interior, U.S. Geological Survey, National Mapping Division, available on the World Wide Web at *http://rockyweb.cr.usgs.gov/nmpstds/demstds.html*, p. 17.

van Horssen, P.W., Pebesma, E.J., and Schot, P.P., 2002, Uncertainties in spatially aggregated predictions from a logistic regression model: Ecological Modelling, v. 154, no. 1–2, p. 93–101.

Veregin, Howard, 1999, Data quality parameters, *in* Longley, P.A., Goodchild, M.F., Maguire, D.J., and Rhind, D.W., eds., Geographical information systems, Volume 1 — Principles and technical issues: New York, John Wiley and Sons, p. 177–189.

Wyss, G.D., and Jorgensen, K.H., 1998, A user's guide to LHS — Sandia's Latin Hypercube Sampling Software: Albuquerque, N. Mex., Sandia National Laboratories, 140 p.

Zhang, Jingxiong, and Goodchild, M.F., 2002, Uncertainty in geographical information: London, UK, Taylor and Francis, 266 p.

Glossary

attribute   As used in this report, an attribute is the property of a geographic object or location in a raster.

cell size   The dimensions in reality that are represented by a single cell in a raster and are measured in map units (Environmental System Research Institute, Inc. [ESRI], 2006).

error   Heuvelink (1998) states that error is the "difference between reality and our representation of reality; it includes not only 'mistakes' or 'faults' but also the statistical concept of 'variation'." Therefore, in the context of data quality, error refers to a measured, observed, calculated, or interpreted value that differs from the true value (Environmental System Research Institute, 2006). In the context of a GIS database, spatial error is from error in position (feature coordinates are wrong) and topology (features do not properly connect, intersect, or adjoin) (Environmental System Research Institute, 2006).

error propagation   The process of error from the input attributes in a GIS operation or model causing **error** and **uncertainty** in the output of the GIS operation or model.

extent   The extent is the minimum bounding rectangle (xmin, ymin, xmax, and ymax) that is defined by coordinate pairs of a data source (Environmental System Research Institute, 2006).

geographic information system (GIS)   A software package that allows users to create, edit, analyze, store, display, and output geographically referenced data.

geoprocessing operation   One or more algorithms that create new geospatial data from geospatial data already held in a GIS database (Krivoruchko and Gotway, 2005). Geoprocessing operations are generally classified as **local**, **neighborhood**, and **global operations** (Environmental System Research Institute, 2006).

geospatial model(ing)   As used in this report, refers to the use of raster data in computational **models** within a **geographic information system** and is synonymous with spatial model(ing).

global operation   An operation that defines or computes an output raster such that each cell location is a function of all the cells in the input rasters (Environmental System Research Institute, 2006).

local operation   An operation that defines or computes a new value (or raster) for a location using input values (or rasters) at the same location (Environmental System Research Institute, 2006).

Map Algebra   Map Algebra (Tomlin, 1990) is a language that defines a syntax for combining map themes by using mathematical operations and analytical functions to create new map themes (Environmental System Research Institute, 2006). In a Map Algebra expression, the operators are a combination of mathematical, logical, or Boolean operator, and spatial analysis functions, and the operands are spatial data and numbers or model coefficients.

metadata   Information that describes content, quality, condition, origin, and many other important characteristics of geospatial data (Environmental System Research Institute, 2006).

model   As used in this report, a model is a simplified and mathematically based representation of reality.

neighborhood operation   An operation that defines or computes new values for locations using the values of other locations within a given distance, direction, or spatial window (Environmental System Research Institute, 2006).

operation   See **geoprocessing operation**.

prediction uncertainty   An expression of confidence about results from a predictive model.

projection   A method that is used to portray the curved surface of the Earth on a flat surface (Environmental System Research Institute, 2006).

Python   An open-source, object-oriented, and high-level programming language with dynamic semantics. For additional details, see the Python Programming Language official Web site on the World Wide Web at: *www.python.org* (accessed November 28, 2008).

raster   A type of geospatial data model that defines space as an array of equally sized cells arranged in rows and columns. Each cell contains an attribute value and location coordinates, which are contained in the data model in the ordering of the matrix. Thus, groups of cells that have the same value represent the same type of geographic feature (Environmental System Research Institute, 2006).

raster processing   A type of **geoprocessing operation** that uses **rasters**.

resolution   The dimensions represented by each cell in a **raster** (Environmental System Research Institute, 2006).

snapping   An automatic editing operation in ArcGIS Desktop that moves points or features within a specified distance of other points or features to coincide exactly with each others' coordinates (Environmental System Research Institute, 2006).

stochastic   A random event or process that can be described by a statistical probability.

uncertainty   An expression of confidence about knowledge (Brown and Heuvelink, 2005; Heuvelink and others, 2007). See also **prediction uncertainty**.

virtual machine   An abstract software implementation of a machine (computer) that executes programs like a real machine. The virtual machine used in REPTool is an abstraction separate from the details of processing the Map Algebra model equation that executes the program that is represented by the Map Algebra model equation.

Appendix 1–Statistical Functions

Algorithm to Compute Inverse Normal Cumulative Distribution Function

The following algorithm was created by Peter J. Acklam (2004) and is used in REPTool to compute the inverse normal cumulative distribution function (CDF) (fig. 27) used for LHS. The Acklam (2004) algorithm is used because the inverse normal cumulative distribution in a nonlinear function that has no closed-form numerical solution. The predicted absolute value from the Acklam (2004) algorithm has a reported relative error of less than 1.15×10^{-9} in the entire region of prediction. Additional details on this algorithm are found at Peter J. Acklam's Web site at: *http://home.online.no/~pjacklam/notes/invnorm/index.html* (accessed December 1, 2008).

The algorithm is computed using input of the probability of the CDF and outputs the corresponding value in the domain for the probability function of variable X. A unique rational approximation is used for the lower, central, and upper regions of the inverse normal CDF. The regions are defined using the probability (p) of the inverse normal cumulative distribution as follows: lower region is $0 < p < 0.02425$; the central region is $0.02425 <= p <= 0.97575$; and the upper region is $0.97575 < p < 1$.

The rational approximation for the lower region ($0 < p < 0.02425$) is defined by

$$y = \frac{\left(\left(\left(\left(\left(c1*q+c2\right)*q+c3\right)*q+c4\right)*q+c5\right)*q+c6\right)}{\left(\left(\left(\left(d1*q+d2\right)*q+d3\right)*q+d4\right)*q+1\right)} \tag{21}$$

and

$$q = \sqrt{-2*\log(p)} \tag{22}$$

where
- y is the value in the domain of the probability function (y-axis, fig. 27),
- $c1$ is a coefficient equal to $-7.784894002430293 \times 10^{-3}$,
- $c2$ is a coefficient equal to $-3.223964580411365 \times 10^{-1}$,
- $c3$ is a coefficient equal to -2.400758277161838,
- $c4$ is a coefficient equal to -2.549732539343734,
- $c5$ is a coefficient equal to 4.374664141464968,
- $c6$ is a coefficient equal to 2.938163982698783,
- $d1$ is a coefficient equal to $7.784695709041462 \times 10^{-3}$,
- $d2$ is a coefficient equal to $3.224671290700398 \times 10^{-1}$,
- $d3$ is a coefficient equal to 2.445134137142996,
- $d4$ is a coefficient equal to 3.754408661907416, and
- p is the cumulative probability value between 0 and 1 (x-axis, fig. 27).

The rational approximation for the central region ($0.02425 <= p <= 0.97575$) is defined by

$$y = \frac{\left(\left(\left(\left(\left(a1*r+a2\right)*r+a3\right)*r+a4\right)*+a5\right)*r+a6\right)*q}{\left(\left(\left(\left(\left(b1*r+b2\right)*r+b3\right)*r+b4\right)*r+b5\right)*r+1\right)} \tag{23}$$

and

$$r = q*q \tag{24}$$

and

$$q = p - 0.5 \tag{25}$$

where
- y is the value in the domain of the probability function (y-axis, fig. 27),
- $a1$ is a coefficient equal to $-3.969683028665376 \times 10^{1}$,
- $a2$ is a coefficient equal to $2.209460984245205 \times 10^{2}$,
- $a3$ is a coefficient equal to $-2.759285104469687 \times 10^{2}$,

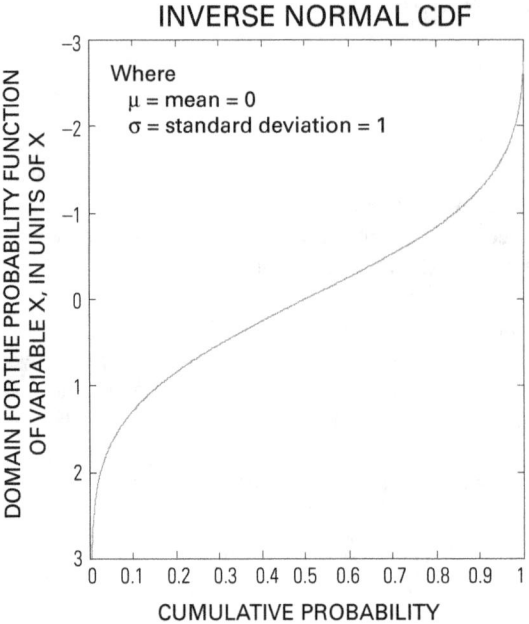

INVERSE NORMAL CDF

Where
μ = mean = 0
σ = standard deviation = 1

Figure 27.   The standard (mean = 0, standard deviation = 1) inverse normal cumulative distribution function (CDF).

$a4$	is a coefficient equal to $1.383577518672690 \times 10^2$,
$a5$	is a coefficient equal to $-3.066479806614716 \times 10^1$,
$a6$	is a coefficient equal to 2.506628277459239,
$b1$	is a coefficient equal to $-5.447609879822406 \times 10^1$,
$b2$	is a coefficient equal to $1.615858368580409 \times 10^2$,
$b3$	is a coefficient equal to $-1.556989798598866 \times 10^2$,
$b4$	is a coefficient equal to $6.680131188771972 \times 10^1$,
$b5$	is a coefficient equal to $-1.328068155288572 \times 10^1$, and
p	is the cumulative probability value between 0 and 1 (x-axis, fig. 27).

The rational approximation for the upper region (0.97575 < p < 1) is defined by

$$y = \frac{-\left(\left(\left(\left(\left(c1 * q + c2\right) * q + c3\right) * q + c4\right) * q + c5\right) * q + c6\right)}{\left(\left(\left(\left(d1 * q + d2\right) * q + d3\right) * q + d4\right) * q + 1\right)} \tag{26}$$

and

$$q = \sqrt{-2 * \log\left(1 - p\right)} \tag{27}$$

where

y	is the value in the domain of the probability function (y-axis, fig. 27),
$c1$	is a coefficient equal to $-7.784894002430293 \times 10^{-3}$,
$c2$	is a coefficient equal to $-3.223964580411365 \times 10^{-1}$,
$c3$	is a coefficient equal to -2.400758277161838,
$c4$	is a coefficient equal to -2.549732539343734,
$c5$	is a coefficient equal to 4.374664141464968,
$c6$	is a coefficient equal to 2.938163982698783,

$d1$ is a coefficient equal to 7.784695709041462 x 10^{-3},
$d2$ is a coefficient equal to 3.224671290700398 x 10^{-1},
$d3$ is a coefficient equal to 2.445134137142996,
$d4$ is a coefficient equal to 3.754408661907416, and
p is the cumulative probability value between 0 and 1 (x-axis, fig. 27).

The Acklam (2004) algorithm refines the lower, central, and upper region approximations using the complementary error function (erfc) and error function (erf). The erfc, where erfc(x) = 1 – erf(x), is appropriate for refining the approximations because there is a relation between the normal CDF and the erfc as follows:

$$\overline{x} = y - \frac{u}{\left(1 + \frac{y * u}{2}\right)} \tag{28}$$

and

$$u = z * \sqrt{2 * \pi} * e^{\frac{y^2}{2}} \tag{29}$$

and

$$z = 0.5 * erfc\left(\frac{-y}{\sqrt{2}}\right) - p \tag{30}$$

where
$\overline{x}$ is the refined value y in the domain of the normal cumulative probability distribution,
y is the value in the domain of the probability function,
π is the mathematical constant Pi that is approximately equal to 3.14159, and
$erfc()$ is the complementary error function, where $erfc() = 1 – erf()$.

Normal Cumulative Distribution Function

The normal cumulative distribution function (CDF), $F(x)$, is expressed as

$$F(x) = \frac{1}{2}\left[1 + erf\left(\frac{x - \mu}{\sigma\sqrt{2}}\right)\right] \tag{31}$$

where
erf is the error function,
x is the value in the domain of the probability function,
    is the mean value of the normal distribution, and
σ is the standard deviation of the normal distribution.

For LHS, it is useful to sample the cumulative probability values (y-axis, fig. 27) relative to the normal CDF rather than sample the actual distribution values (x-axis, fig. 27) because the nonoverlapping intervals are defined by quantiles of equal probability that are expressed as cumulative probability units from 0 to 1 (y-axis, fig. 27) (see the section "**Latin Hypercube Sampling Method**"). The actual distribution values, x, (x-axis, fig. 27) are calculated by inverting the normal CDF using the Acklam (2004) algorithm (see the section "**Algorithm to Compute Inverse Normal Cumulative Distribution Function**") as follows:

$$x = \mu + \sigma * \overline{x} \tag{32}$$

where
    is the mean value of the normal distribution,
σ is the standard deviation of the normal distribution, and
$\overline{x}$ is the refined value y in the domain of the normal cumulative probability distribution (see equation 26).

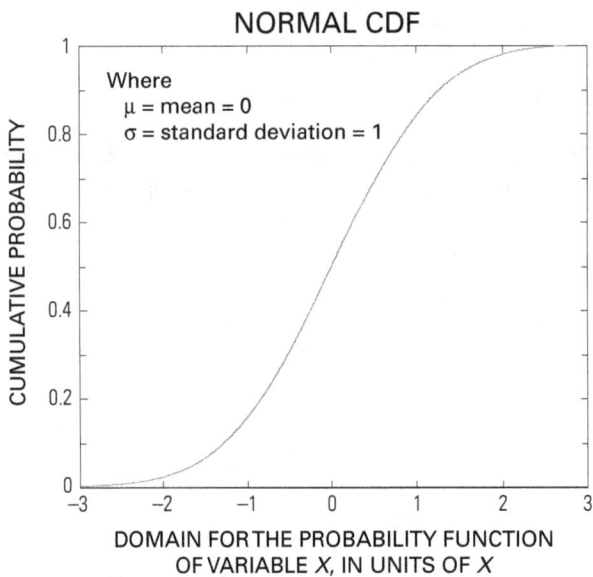

Figure 28.   The standard (mean = 0, standard deviation = 1) normal cumulative distribution function (CDF).

Lognormal Cumulative Distribution Function

The lognormal cumulative distribution function (CDF), $F(x)$, is expressed as

$$F(x) = \frac{1}{2} + \frac{1}{2}\, erf\left(\frac{\ln(x) - \mu}{\sigma\sqrt{2}}\right)$$ (33)

where

erf	is the error function,
$\ln$	is the natural logarithm,
x	is the value in the domain of the probability function,
	is the mean value of the lognormal distribution, and
σ	is the standard deviation of the lognormal distribution.

Similar to the normal CDF, it is useful to sample the cumulative-probability values relative to the normal CDF rather than sample the actual distribution values during LHS (see "**Normal Cumulative Distribution Function**"). The actual lognormal distribution values, x, (x-axis, fig. 29) are calculated by inverting the normal CDF using the Acklam (2004) algorithm as follows:

$$x = e^{\left(\mu + 0*\bar{x}\right)}$$ (34)

where

e	is the base of the natural logarithm,
	is the mean value of the lognormal distribution,
σ	is the standard deviation of the lognormal distribution, and
$\bar{x}$	is the refined value y in the domain of the normal cumulative probability distribution (see equation 26).

Uniform Cumulative Distribution Function

The uniform cumulative distribution function (CDF), $F(x)$, is expressed as

$$F(x) = \frac{x - a}{b - a} \qquad (35)$$

where

x	is the value in the domain of the probability function of variable x,
a	is the minimum value of the uniform distribution, and
b	is the maximum value of the uniform distribution.

For LHS, it is useful to rearrange equation 33 and use the cumulative-probability value, $F(x)$, as input to solve for x as follows:

$$x = F(x)(b - a) + a \qquad (36)$$

where

x	is the value in the domain of the probability function of variable x,
$F(x)$	is the cumulative probability value between 0 and 1 for the uniform CDF,
a	is the minimum value of the uniform distribution, and
b	is the maximum value of the uniform distribution.

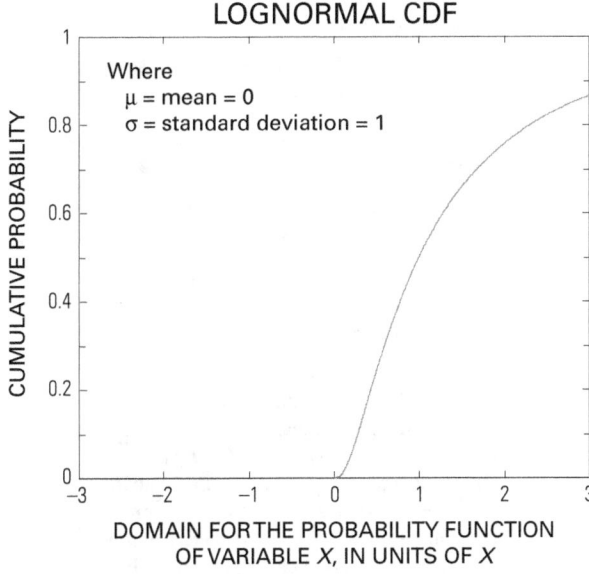

Figure 29. The standard (mean = 0, standard deviation = 1) lognormal cumulative distribution function (CDF).

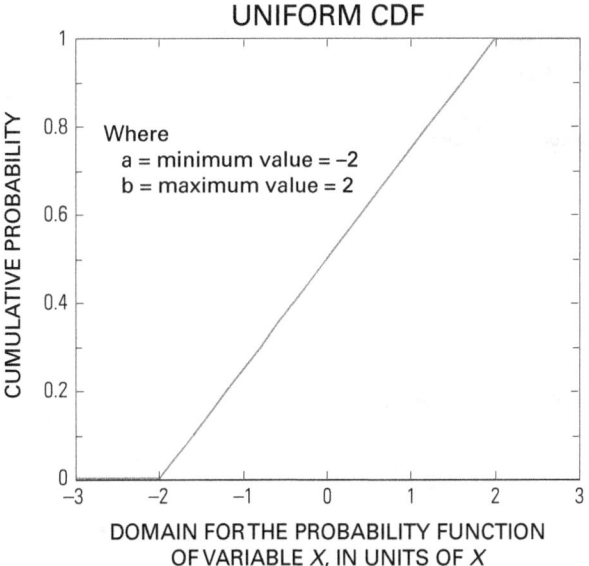

Figure 30. Uniform cumulative distribution function (CDF) for a hypothetical variable x that has a minimum value (a) of –2 and maximum value (b) of 2.

Appendix 2–Command-Line Syntax and Python Scripting

Appendix 2 provides command-line syntax and Python scripting use and examples for REPTool. The information provided in Appendix 2 also is available in the REPTool Help page.

Command-Line Syntax

reptool_custom <Input_rasters;Input_rasters...> {Spatially_variable_error} {Input_spatially_variable_error_rasters;Input_spatially_variable_error_rasters...} {Input_errors} {Model_uses_coefficients_} {Specify_coefficient_file} <Distribution_type> {Model_equation} {Model_equation_text_file} <10 | 25 | 50 | 100> {Output_percentiles} {Calculate_relative_variance_contribution__RVC_} <Output_workspace>

Parameters

Expression	Explanation
<Input_rasters;Input_rasters...>	The rasters used as input for the model.
{Spatially_variable_error}	Command-line syntax if the error associated with any one of the input rasters varies spatially.
{Input_spatially_variable_error_rasters;Input_spatially_variable_error_rasters...}	The error raster data sets associated with each input raster data set.
{Input_errors}	Input error (as percent) for each of the input raster data sets. The error values must be comma separated and be in the same input order as the input raster data sets.
{Model_uses_coefficients_}	Command-line syntax if the model equation contains any coefficients.
{Specify_coefficient_file}	The text file that contains the information about the coefficients and associated errors.
<Distribution_type>	The type of error distribution to be used for each of the input data sets.
{Model_equation}	The model equation containing the coefficients, variables (input rasters), operators, and functions that will be used to process the input data sets.
{Model_equation_text_file}	The text file that contains the model equation. Note – The Model equation text file option must be used to input the model equation if REPTool is run from the ArcMap command line. The ArcMap command-line interpreter does not allow for special characters in model equations, which will cause an error if the model equation is specified using the Model-equation option. Therefore, it is necessary to input the model equation as a text file when running REPTool from the ArcMap command line.
<10	25	50	100>	The number of iterations used to define the probability distribution for the Latin Hypercube Sampling (LHS) technique.
{Output_percentiles}	The percentiles (semicolon separated) of the final output distribution surrounding the model results.

{Calculate_relative_variance_contribution__RVC_}	Command-line syntax to calculate the relative variance contributions (RVC) of model error versus the RVC of the input data sets (explanatory variables).
<Output_workspace>	Output workspace where results will be written and temp files will be processed.

Command-Line Example

Reptool_custom D:\MyData\irrpct3;D:\MyData\soilinf3 # # 10,20 true D:\MyData\coeff.txt Normal # D:\MyData\equation.txt 10 5;90 true D:\MyData\finalmap2

Note — the '#' is used as a placeholder for optional parameters.

Scripting Syntax

Reptool_custom (Input_rasters, Spatially_variable_error, Input_spatially_variable_error_rasters, Input_errors, Model_uses_coefficients_, Specify_coefficient_file, Distribution_type, Model_equation, Model_equation_text_file, Number_of_interations, Output_percentiles, Calculate_relative_variance_contribution__RVC_, Output_workspace)

Parameters

Expression	Explanation
Input_rasters (Required)	The rasters used as input for the model.
Spatially_variable_error? (Optional)	Scripting syntax if the error associated with any one of the input rasters varies spatially.
Input_spatially_variable_error rasters (Optional)	The error raster data sets error associated with each input raster data set.
Input_errors (Optional)	Input errors (as percent) for each of the input raster data sets. The error values must be comma separated and be in the same input order as the input raster data sets.
Model_uses_coefficients? (Optional)	Scripting syntax if your model equation contains any coefficients.
Specify_coefficient_file (Optional)	The text file that contains the information about the coefficients and associated errors.
Distribution_type (Required)	The type of error distribution to be used for each of the input data sets.
Model-equation (Optional)	The model equation containing the coefficients, variables (input rasters), operators, and functions that will be used to process the input data sets.
Model_equation_text-file (Optional)	The text file that contains the model equation. Note – The Model equation text file option must be used to input the model equation if REPTool is run from the ArcMap command line. The ArcMap command-line interpreter does not allow for special characters in model equations, which will cause an error if the model equation is specified using the Model-equation option. Therefore, it is necessary to input the model equation as a text file when running REPTool from the ArcMap command line.
Number_of_iterations (Required)	The number of iterations used to define the probability distribution for the Latin Hypercube Sampling (LHS) technique.
Output_percentiles (Optional)	The percentiles (semicolon separated) of the final output distribution surrounding the model results.
Calculate_relative_variance_contribution (RVC) (Optional)	Scripting syntax to calculate the relative variance contributions (RVC) of model error compared to the RVC of the input data sets (explanatory variables).

Output_workspace (Required)	Output workspace where results will be written and temp files will be processed.

Script Example

```
# reptool.py
# Description:
#  Quantify error propagation and uncertainty during raster processing
# Requirements: None
# Date: July 21, 2008
# Import system modules
import arcgisscripting
# Create the Geoprocessor object
gp = arcgisscripting.create()
try:
        # Set local variables
                    InRasters = "D:/REPTool_v_1_0/Example/dtw;D:/ REPTool_v_1_0/Example /aqbottom"
            InErrors = "10,20"
            coef_file = "D:/ REPTool_v_1_0/Example /coef_file.txt"
            InEq = "var01 - var00"
            Dist = "Normal"
            no_itns = "10"
            out_pctls = "5;25;75;95"
            out_wsp = "D:/ REPTool_v_1_0/Example /Analysis"
            # Check out Spatial Analyst extension license
            gp.CheckOutExtension("Spatial")
            # Run REPTool
                    gp reptool_custom(InRasters,#,#,InErrors,"true",coef_file,Dist,InEq,#,no_itns,out_pctls,"true",out_wsp)
except:
        # If an error occurred while running a tool, then print the messages.
        print gp.GetMessages()
```

Appendix 3 – Developer Documentation

Overview

Appendix 3 presents an overview of REPTool version 1.0 Python software-package architecture. The purpose of the overview is to provide information for software developers seeking details of the REPTool implementation; this information is not needed for general REPTool use. Appendix 3 outlines a general overview of each package model. The contents of Appendix 3 are listed below. The reader is encouraged to begin with **How To Read Developer Documentation**.

Contents

How to Read Developer Documentation

Python Package

Python Class
-Class Attribute 1
-Class Attribute 2
-|
-|
-Class Attribute N
+method_1_()
+method_2_()
+|()
+|()
+method_N_()

Python Module
-Module Attribute 1
-Module Attribute 2
-|
-|
-Module Attribute N
+method_1_()
+method_2_()
+|()
+|()
+method_N_()

A Python Class: A
-A's Attribute 1
-A's Attribute 2
-|
-|
-A's Attribute N
+A's method_1_()
+A's method_2_()
+|()
+|()
+A's method_N_()

B inherits from A

A Python Class: B
-A's Attribute 1
-A's Attribute 2
-|
-|
-A's Attribute N
-B's Attribute 1
-B's Attribute 2
-|
-|
-B's Attribute N
+A's method_1_()
+A's method_2_()
+|()
+|()
+A's method_N_()
+B's method_1_()
+B's method_2_()
+|()
+|()
+B's method_N_()

A Note Shape

Models describing REPTool's Python-package architecture use Microsoft Visio Shapes to describe elements of a Python package. The Shape notation generally conforms to Unified Modeling Language (UML) standards. Shapes in each model page describe Python modules and Python classes. The Note Shape, like the one containing this text, uses a light gray background color and is linked to other Shape symbols by a dotted line. Python-package elements are described by the Notes linked to their Shapes.

The Shape describing a Python-package name uses a light blue background and pages describing a single package place this Shape near the upper left corner. Multi-page models append a designator (1 of 6) to a package name. Python modules and classes described in the models use the same Shape; however, module Shapes use a medium gray background color and class Shapes use a white background color. Examples of module and class Shapes and notation are shown in "How To Read Developer Documentation."

For all Shapes describing a Python class the module name for the class described is the same as the class name. Shapes describing Python modules and classes generally omit descriptions of module and class Attributes because the information is not relevant for this architecture overview. The __init__() method a Python class uses as the default constructor is assumed for all class descriptions. Notation for elements in a Shape with multiple instances of generic names (that is err00, err01, err03) are specified by following the generic name with <##> or <###> to indicate the Python package element described by a Shape defines more than one instance of the generic name with 2 or 3 digits appended to each case, respectively. In a few cases, a Note Shape describes a Python package element simply as INACTIVE, indicating that the element was created for test\ or development scenarios, but is not active during execution for the current version of REPTool (1.0). The INACTIVE elements may be used by future versions of REPTool.

The control package and certain classes in other packages are generalized classes that are extended by other classes. The rules of inheritance for Python classes are similar to other object-oriented languages, such as C++ and Java. As depicted in "How To Read Developer Documentation," A Python Class: B inherits from A Python Class: A. The Python rule for inheritance generally states that the attributes and methods of the parent class, A Python Class: A in the example case, are automatically shared by the child class, A Python Class: B.

A Model Shape
(See examples above.)

System Architecture

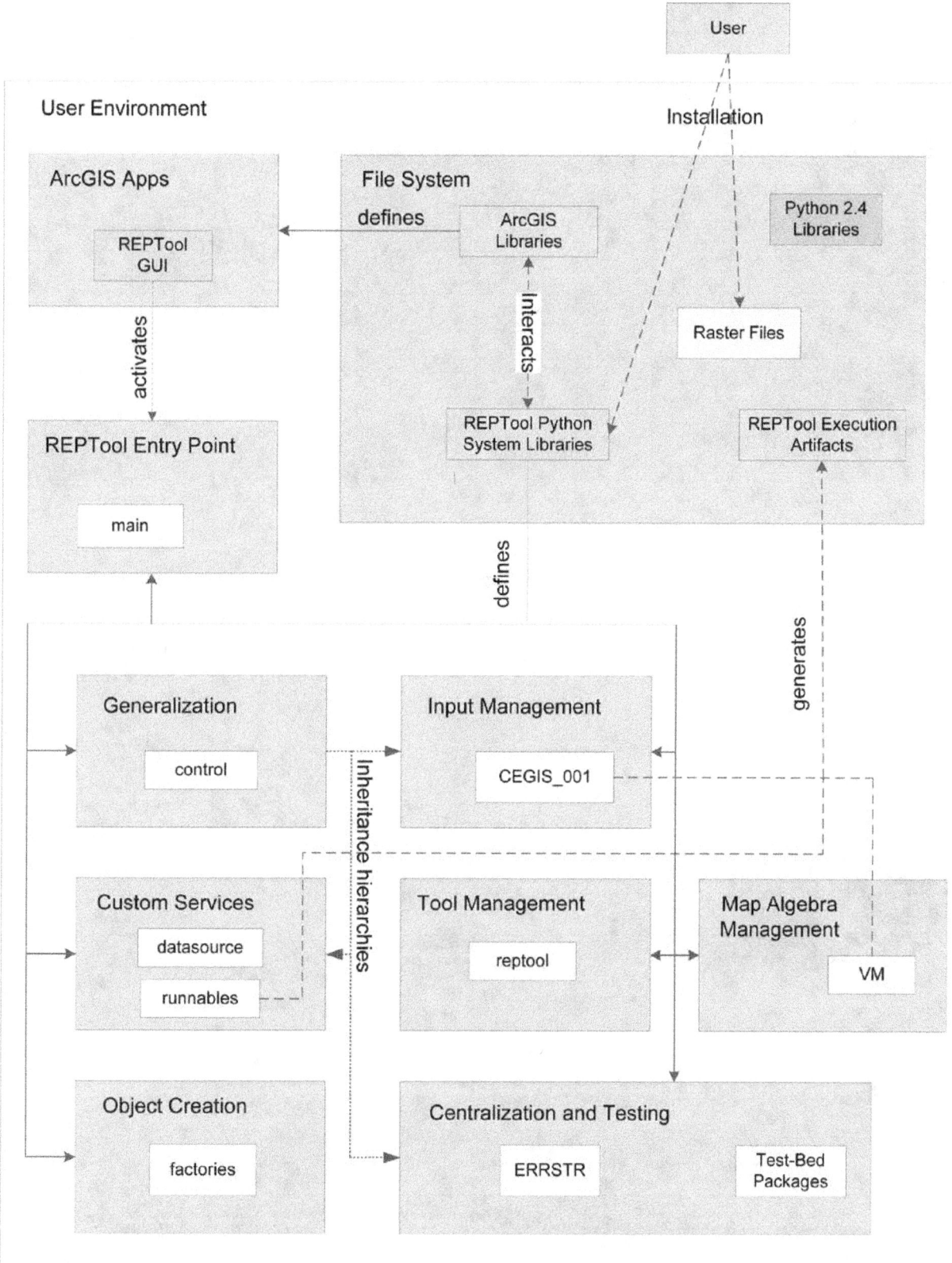

Package Architecture: CEGIS_001

The aprocess class generalizes processing the parameters returned from a GUI interface.

This central grouping of classes inherits from aprocess and performs the verification and validation and subsequent processing on various GUI parameter types. For example, boolchk confirms a string value representing the value of a gui checkbox is either 'true' or 'false' and returns the Python boolean equivalent.

The iterprocess class and multiprocess class inherit from aprocess and generically operate by iterating calls to the process() method of an instance of class aprocess or a child of class aprocess or by chaining of input and output to the process() methods for a list containing instances of aprocess class or aprocess child classes.

The __cons_01__ module defines a set of GUI constants and processor set constants and several accessor methods.

The gp_utl module contains a few utility methods for interacting with ArcGIS libraries.

The mgr_input_001 class is a child of control.mgr_input and manages tasking for input verification and validation and the subsequent setup work for data processing, including the compilation of the map algebra equation and creation of a centralized set of objects used throughout linear execution by REPTool (1.0).

The __NSP00__, utl_01, and errstr_000 modules centralize error generation for the CEGIS_001 package. The errstr_000 module actually resides under the ERRSTR package in the architecture for storage purposes, but is functionally a member of CEGIS_001. The errstr_<###> modules store message list structures defined for a package. The err<##> attributes store message list structures for a class by method. An ERRSTR attribute centralizes the err<##> attributes in a list data structure for access by a namespace.

The __NSP<##>__ modules provide interfaces between namespace utility modules and error string modules. The M<###> attributes store method name keys by class for a given package. A METHODS attribute centralizes the M<###> instances in a list data structure for accessor methods. The utl_<##> modules provide an interface between a namespace and a class implementation. The CN() method performs classname processing and formatting. The EXC() method accesses centralization for a method's exception creation needs and returns a formatted exception to the caller. The getCNI() and getM() methods forward calls to a namespace interface.

Package Architecture: control

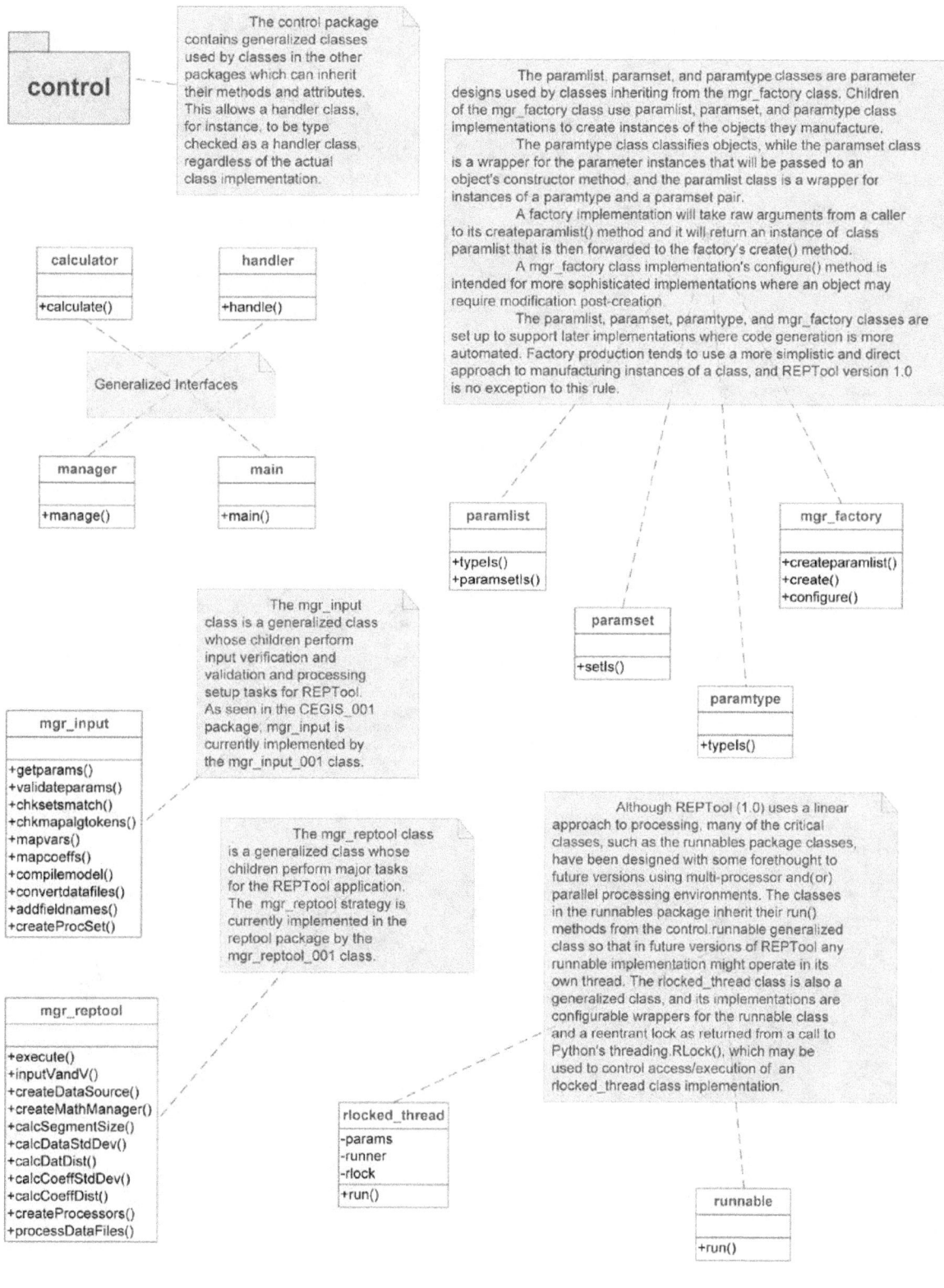

control

The control package contains generalized classes used by classes in the other packages which can inherit their methods and attributes. This allows a handler class, for instance, to be type checked as a handler class, regardless of the actual class implementation.

The paramlist, paramset, and paramtype classes are parameter designs used by classes inheriting from the mgr_factory class. Children of the mgr_factory class use paramlist, paramset, and paramtype class implementations to create instances of the objects they manufacture.

The paramtype class classifies objects, while the paramset class is a wrapper for the parameter instances that will be passed to an object's constructor method, and the paramlist class is a wrapper for instances of a paramtype and a paramset pair.

A factory implementation will take raw arguments from a caller to its createparamlist() method and it will return an instance of class paramlist that is then forwarded to the factory's create() method.

A mgr_factory class implementation's configure() method is intended for more sophisticated implementations where an object may require modification post-creation.

The paramlist, paramset, paramtype, and mgr_factory classes are set up to support later implementations where code generation is more automated. Factory production tends to use a more simplistic and direct approach to manufacturing instances of a class, and REPTool version 1.0 is no exception to this rule.

calculator

+calculate()

handler

+handle()

Generalized Interfaces

manager

+manage()

main

+main()

paramlist

+typeIs()
+paramsetIs()

mgr_factory

+createparamlist()
+create()
+configure()

paramset

+setIs()

paramtype

+typeIs()

The mgr_input class is a generalized class whose children perform input verification and validation and processing setup tasks for REPTool. As seen in the CEGIS_001 package, mgr_input is currently implemented by the mgr_input_001 class.

mgr_input

+getparams()
+validateparams()
+chksetsmatch()
+chkmapalgtokens()
+mapvars()
+mapcoeffs()
+compilemodel()
+convertdatafiles()
+addfieldnames()
+createProcSet()

The mgr_reptool class is a generalized class whose children perform major tasks for the REPTool application. The mgr_reptool strategy is currently implemented in the reptool package by the mgr_reptool_001 class.

Although REPTool (1.0) uses a linear approach to processing, many of the critical classes, such as the runnables package classes, have been designed with some forethought to future versions using multi-processor and(or) parallel processing environments. The classes in the runnables package inherit their run() methods from the control.runnable generalized class so that in future versions of REPTool any runnable implementation might operate in its own thread. The rlocked_thread class is also a generalized class, and its implementations are configurable wrappers for the runnable class and a reentrant lock as returned from a call to Python's threading.RLock(), which may be used to control access/execution of an rlocked_thread class implementation.

mgr_reptool

+execute()
+inputVandV()
+createDataSource()
+createMathManager()
+calcSegmentSize()
+calcDataStdDev()
+calcDatDist()
+calcCoeffStdDev()
+calcCoeffDist()
+createProcessors()
+processDataFiles()

rlocked_thread

-params
-runner
-rlock

+run()

runnable

+run()

Package Architecture: datasources

The datasources in this package all inherit from VM.mapalg_datasource.mapalg_datasource.

The datastddevrunnable_datasource_000 class performs standard deviation calculations for each of the data points in an input raster file.

datastddevrunnable_datasource_000

+__getVar__()
+__getLit__()
+__next__()
+__reset__()
+__getCursors__()

The datadistrunnable_datasource_000 class is initialized with a list containing a nametag associated with a raster file by the application, such as 'var00', and the associated nametag for the corresponding error variable, 'err00'. The list also contains the processor set created during input verification and validation and the primary datasource linked to the data files specified by the user in the application GUI. Multiple instances of this class are created, one for each variable associated with a data raster, and used by the calculators, which generate a distribution around each data point.

datadistrunnable_datasource_000

+__getVar__()
+__getLit__()
+__next__()
+__reset__()
+__getCursors__()
+readSigma()
+readChar()

The mapalgrunnable_datasource_000 class is a service provider for the _001 version of the same name. Initialized with a datasource linked to the raster files and the processor set created during input verification and validation, it performs the reading and copying of a distribution set associated with a data point when its __next__() method is invoked. Other requests, not involving a distribution set, are passed through to the datasource given to this class at initialization. This class represents a service provider for the other datasource classes involved in map algebra calculations.

mapalgrunnable_datasource_000

+copy()
+__getVar__()
+__getLit__()
+__next__()
+__reset__()
+__getCursors__()
+readDist()

Like the _000 version, the mapalgrunnable_datasource_001 class represents a service provider for the datasource classes named vm_datasource_<###>. Initialized with the processor set created during input verification and validation, this class will pass calls not involving a distribution set to the datasource linked to the raster files. Multiple instances of this class service the datasources that service the virtual machine instances, one for each virtual machine datasource.

The mapalgrunnable_datasource_001 class will get a distribution set associated with a data point in a raster from the _000 version of this class. At each execution of a virtual machine, perfoming a map algebra function, instances of vm_datasource_<###> will fill values for the variables, coefficients, constants, and literals used by the map algebra equation specified by the user, using values provided by this class. For a distribution set, a call to __next__() will advance the value provided to the next value in the set until all values in the set have been used. The next set is then loaded from the _000 service provider.

mapalgrunnable_datasource_001

+__getVar__()
+__getLit__()
+__next__()
+__reset__()
+__getCursors__()

The three implementations of vm_datasource_<###> class in the initial version of REPTool service instances of VM.vm.vm, virtual machines that execute the map algebra equation compiled from the user's input. The _000 version provides data from distribution sets for both coefficients and variables (raster data). The _001 version provides data from a distribution set for variables only, and the _002 version provides data from a distribution set for coefficients only, and all three use mapalgrunnable_datasource_001 as a service provider. The three lines of execution provide data for the RVC calculations. This package is a primary site for reimplementation in future multiprocessing versions.

vm_datasource_000

+__getVar__()
+__getLit__()
+__next__()
+__reset__()
+__getCursors__()

vm_datasource_001

+__getVar__()
+__getLit__()
+__next__()
+__reset__()
+__getCursors__()

vm_datasource_002

+__getVar__()
+__getLit__()
+__next__()
+__reset__()
+__getCursors__()

Development Test-Bed Packages

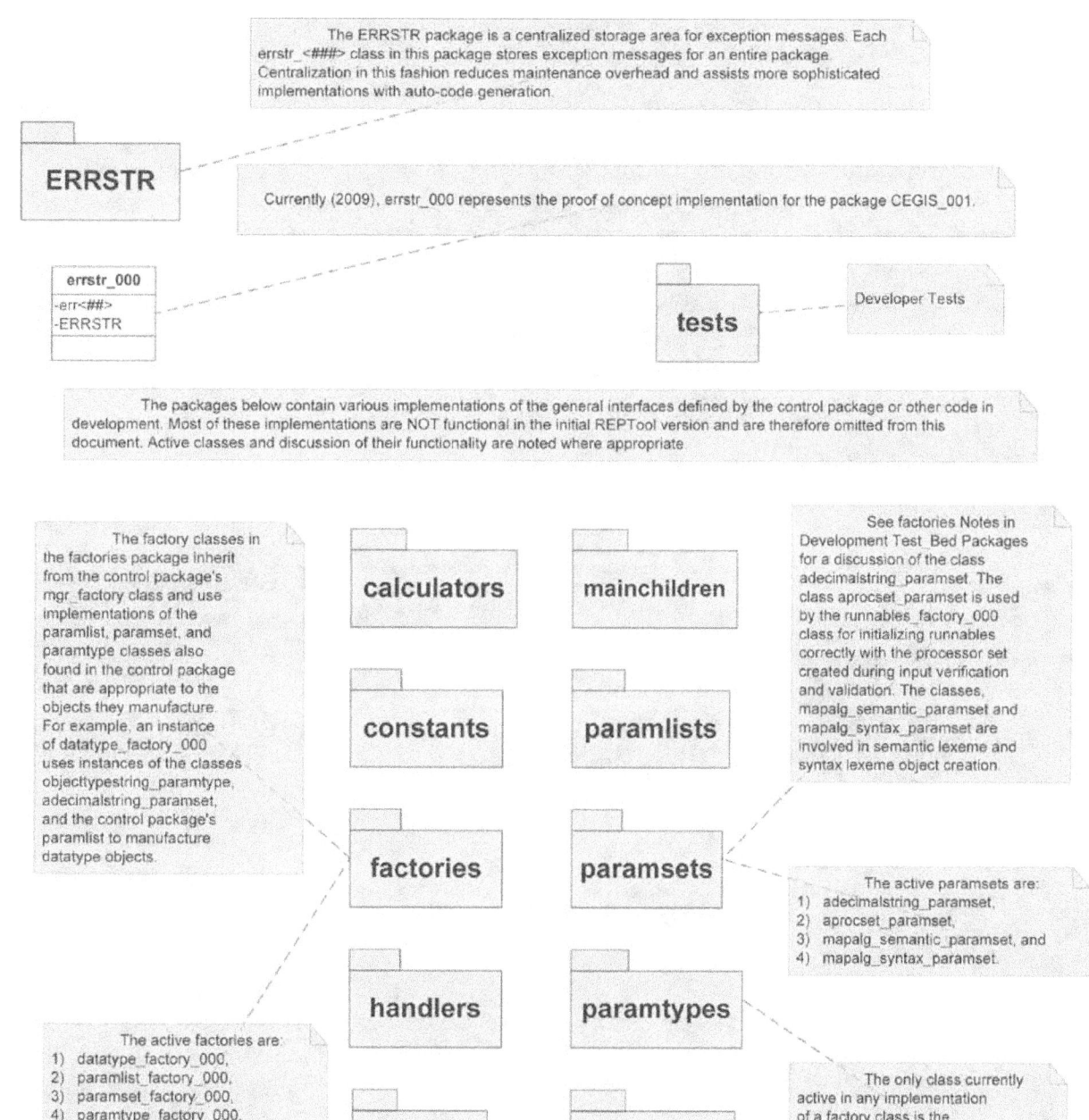

The ERRSTR package is a centralized storage area for exception messages. Each errstr_<###> class in this package stores exception messages for an entire package. Centralization in this fashion reduces maintenance overhead and assists more sophisticated implementations with auto-code generation.

ERRSTR

Currently (2009), errstr_000 represents the proof of concept implementation for the package CEGIS_001.

errstr_000
-err<##>
-ERRSTR

tests

Developer Tests

The packages below contain various implementations of the general interfaces defined by the control package or other code in development. Most of these implementations are NOT functional in the initial REPTool version and are therefore omitted from this document. Active classes and discussion of their functionality are noted where appropriate.

The factory classes in the factories package inherit from the control package's mgr_factory class and use implementations of the paramlist, paramset, and paramtype classes also found in the control package that are appropriate to the objects they manufacture. For example, an instance of datatype_factory_000 uses instances of the classes objecttypestring_paramtype, adecimalstring_paramset, and the control package's paramlist to manufacture datatype objects.

calculators

mainchildren

See factories Notes in Development Test_Bed Packages for a discussion of the class adecimalstring_paramset. The class aprocset_paramset is used by the runnables_factory_000 class for initializing runnables correctly with the processor set created during input verification and validation. The classes, mapalg_semantic_paramset and mapalg_syntax_paramset are involved in semantic lexeme and syntax lexeme object creation.

constants

paramlists

factories

paramsets

The active paramsets are:
1) adecimalstring_paramset,
2) aprocset_paramset,
3) mapalg_semantic_paramset, and
4) mapalg_syntax_paramset.

handlers

paramtypes

The active factories are:
1) datatype_factory_000,
2) paramlist_factory_000,
3) paramset_factory_000,
4) paramtype_factory_000,
5) runnable_factory_000,
6) semlexeme_factory_000, and
7) synlexeme_factory_000.

managers

prototypes

The only class currently active in any implementation of a factory class is the objecttypestring_paramtype class and it serves as a general purpose proof of concept.

Package Architecture: main

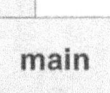

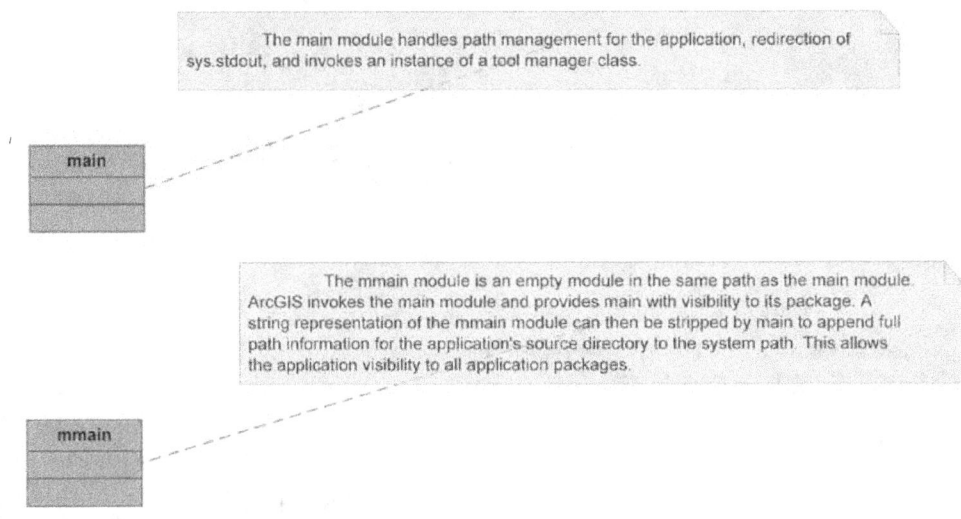

The main module handles path management for the application, redirection of sys.stdout, and invokes an instance of a tool manager class.

main

The mmain module is an empty module in the same path as the main module. ArcGIS invokes the main module and provides main with visibility to its package. A string representation of the mmain module can then be stripped by main to append full path information for the application's source directory to the system path. This allows the application visibility to all application packages.

mmain

Package Architecture: reptool

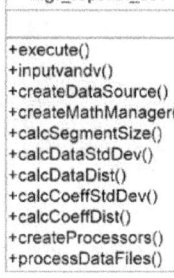

reptool

The mgr_reptool_000 class represents a development strategy test-bed that is not currently implemented.

mgr_reptool_000
+ __isconfigured__()
+ __loadconstants__()
+ __loadfactories__()
+ __loadstrategies__()
+ __main__()
+ __mainstrategy__()
+ __handle__()
+ __cleanup__()
+ __exit__()

The mgr_reptool_001 class represents the current working version of a tool manager class for the REPTool application. The class is responsible for controlling the workflow for the application. Other managers, such as the mgr_input_001 class, are invoked by mgr_reptool_001 to handle tasks like input verification and validation and completed work segments are forwarded between strategy handlers by mgr_reptool_001. For example, the processor set created during input verification and validation is forwarded to the handlers instantiating calculation runnables and datasources. The term handlers is used here in the generic sense, and not in the sense that handler necessarily indicates an implementation of the control handler class.

mgr_reptool_001
+execute()
+inputvandv()
+createDataSource()
+createMathManager()
+calcSegmentSize()
+calcDataStdDev()
+calcDataDist()
+calcCoeffStdDev()
+calcCoeffDist()
+createProcessors()
+processDataFiles()

Package Architecture: runnables 1 of 2

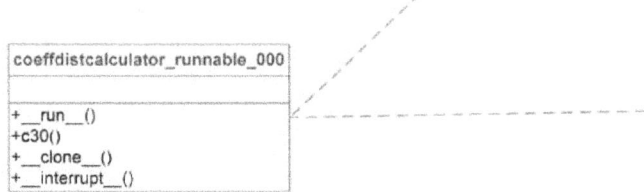

runnables 1 of 2

The coeffdistcalculator_runnable_000 class represents the working version of a strategy to calculate distribution values for coefficients specified by the user in REPTool's GUI. The clone and interrupt methods are development strategies not implemented, and the c30 method is a developer utility. The primary _run_() method defined by the control.runnable class and implemented in coeffdistcalculator_runnable_000 manages the strategy. The class is initialized with an application variable name for a coefficient. The processor set is created during input verification and validation and an instance of a math manager. Coefficient values and calculation results are centralized in the processor set to avoid IO overhead whenever possible. The implementation handles interaction between calculation needs and the math manager.

coeffdistcalculator_runnable_000
+__run__()
+c30()
+__clone__()
+__interrupt__()

*** A Side Note ***

The coefficient standard deviation calculation requirements are trivial when compared to other calculation requirements handled by the runnables package and are therefore handled by the mgr_reptool_001 class directly.

The datastddevcalculator_runnable_000 class represents the working version of a strategy for calculating standard deviations for raster input files. Instances of datastddevcalculator_runnable_000 are provided with a variable name associated with an input raster file, a data source, and an instance of a math manager. An instance of datastddevcalculator_runnable_000 is designed to calculate for a single application variable name that is associated with a raster input file specified by the user in the application GUI. For each application variable name, an instance of datastddevcalculator_runnable_000 is created to run its calculations. In the current (2009) version of the application, instances of the class datastddevcalculator_runnable_000 operate consecutively. In more sophisticated versions, a redesign of the data-source architecture acting as service providers for these runnables will allow implementations to be run in their own threads.

datastddevcalculator_runnable_000
+reset()
+__run__()
+clearinvalids()
+__clone__()
+__interrupt__()

Package Architecture: runnables 2 of 2

runnables 2 of 2

The datadistcalculator_runnable_000 class represents the working version of a strategy for calculating distribution sets for raster input data points. Like the runnable implementation for coefficient distribution set calculations, datadistcalculator_runnable_000 is instantiated with a variable name, a data source, and a math manager, but it is also handed the processor set created during input verification and validation. Strategy management and design architecture in the datadistcalculator_runnable_000 class follow the same pattern found in the other runnables for future development requirements.

datadistcalculator_runnable_000
+reset()
+__run__()
+clearinvalids()
+stdcalcs01()
+stdcalcs02()
+normal_dist_calcs()
+lognormal_dist_calcs()
+uniform_dist_calcs()
+store()
+__clone__()
+__interrupt__()

The mapalgcalculator_runnable_000 class represents the working version of a strategy for implementing the map algebra equation specified by the user in the application GUI, using data from the input raster files and coefficients also specified by the user. Strategy management and architecture implemented by mapalgcalculator_runnable_000 mimic those runnables that manage calculations leading up to the map algebra calculations, and class instantiation takes a data source, a math manager, and the processor set created during input verification and validation. However, mapalgcalculator_runnable_000 uses the VM package architecture to do the actual processing, and the data source architecture contains several layers of service providers.

Percentile and RVC calculations are also bundled into mapalgcalculator_runnable_000 to avoid the input/output overhead that would be required to separate these operations. Likewise, final output calculations and persistence management are also handled by the mapalgcalculator_runnable_000 class.

mapalgcalculator_runnable_000
+initperc()
+resetperc()
+initrvc()
+resetrvc()
+reset()
+__run__()
+checknext()
+clearinvalids()
+handleinvalid()
+sort()
+store()
+calcrvc()
+calcperc()
+__clone__()
+__interrupt__()

Package Architecture: Virtual Machine (VM) 1 of 6

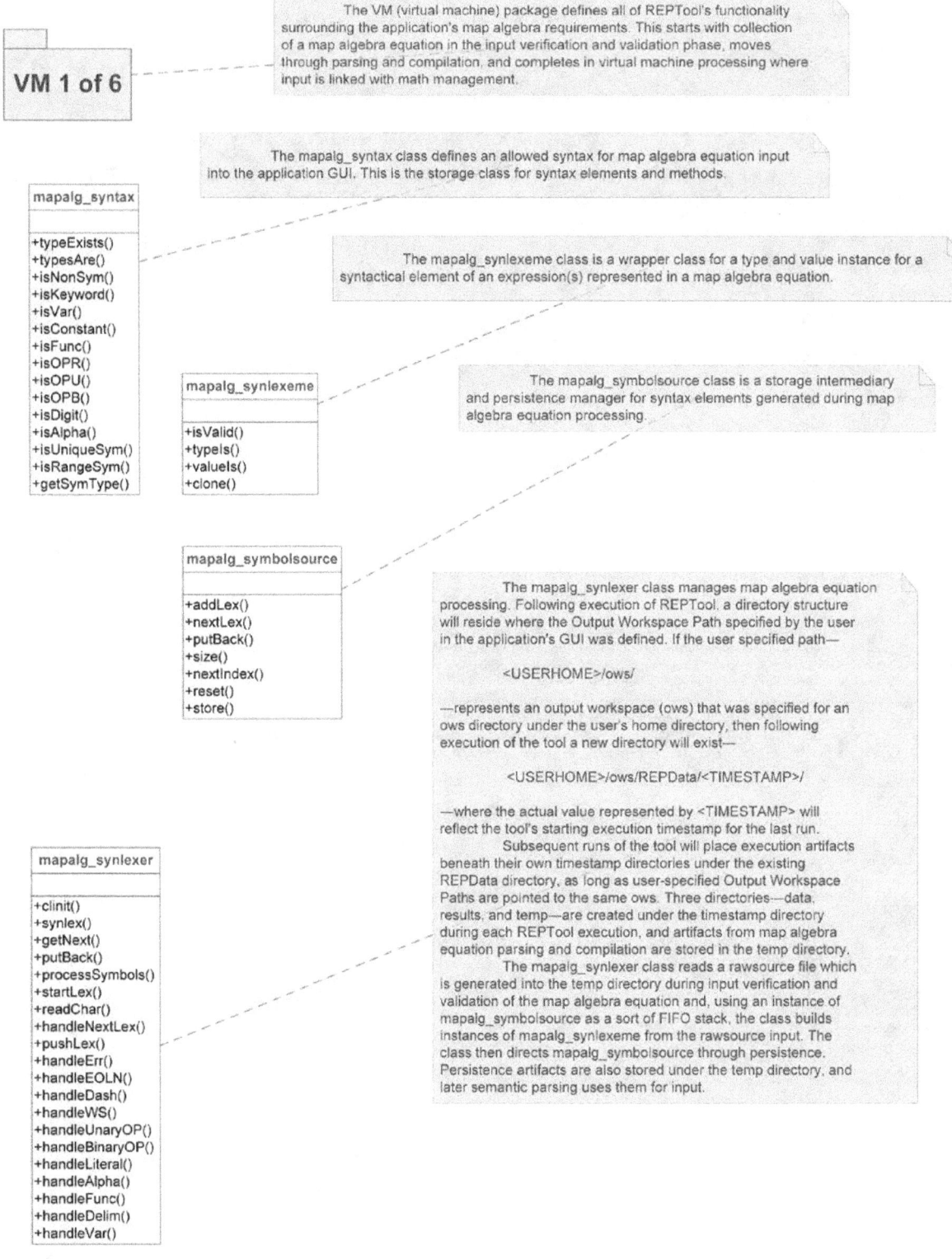

VM 1 of 6

The VM (virtual machine) package defines all of REPTool's functionality surrounding the application's map algebra requirements. This starts with collection of a map algebra equation in the input verification and validation phase, moves through parsing and compilation, and completes in virtual machine processing where input is linked with math management.

The mapalg_syntax class defines an allowed syntax for map algebra equation input into the application GUI. This is the storage class for syntax elements and methods.

mapalg_syntax

+typeExists()
+typesAre()
+isNonSym()
+isKeyword()
+isVar()
+isConstant()
+isFunc()
+isOPR()
+isOPU()
+isOPB()
+isDigit()
+isAlpha()
+isUniqueSym()
+isRangeSym()
+getSymType()

The mapalg_synlexeme class is a wrapper class for a type and value instance for a syntactical element of an expression(s) represented in a map algebra equation.

The mapalg_symbolsource class is a storage intermediary and persistence manager for syntax elements generated during map algebra equation processing.

mapalg_synlexeme

+isValid()
+typeIs()
+valueIs()
+clone()

mapalg_symbolsource

+addLex()
+nextLex()
+putBack()
+size()
+nextIndex()
+reset()
+store()

The mapalg_synlexer class manages map algebra equation processing. Following execution of REPTool, a directory structure will reside where the Output Workspace Path specified by the user in the application's GUI was defined. If the user specified path—

<USERHOME>/ows/

—represents an output workspace (ows) that was specified for an ows directory under the user's home directory, then following execution of the tool a new directory will exist—

<USERHOME>/ows/REPData/<TIMESTAMP>/

—where the actual value represented by <TIMESTAMP> will reflect the tool's starting execution timestamp for the last run. Subsequent runs of the tool will place execution artifacts beneath their own timestamp directories under the existing REPData directory, as long as user-specified Output Workspace Paths are pointed to the same ows. Three directories—data, results, and temp—are created under the timestamp directory during each REPTool execution, and artifacts from map algebra equation parsing and compilation are stored in the temp directory.

The mapalg_synlexer class reads a rawsource file which is generated into the temp directory during input verification and validation of the map algebra equation, and, using an instance of mapalg_symbolsource as a sort of FIFO stack, the class builds instances of mapalg_synlexeme from the rawsource input. The class then directs mapalg_symbolsource through persistence. Persistence artifacts are also stored under the temp directory, and later semantic parsing uses them for input.

mapalg_synlexer

+clinit()
+synlex()
+getNext()
+putBack()
+processSymbols()
+startLex()
+readChar()
+handleNextLex()
+pushLex()
+handleErr()
+handleEOLN()
+handleDash()
+handleWS()
+handleUnaryOP()
+handleBinaryOP()
+handleLiteral()
+handleAlpha()
+handleFunc()
+handleDelim()
+handleVar()

Package Architecture: Virtual Machine (VM) 2 of 6

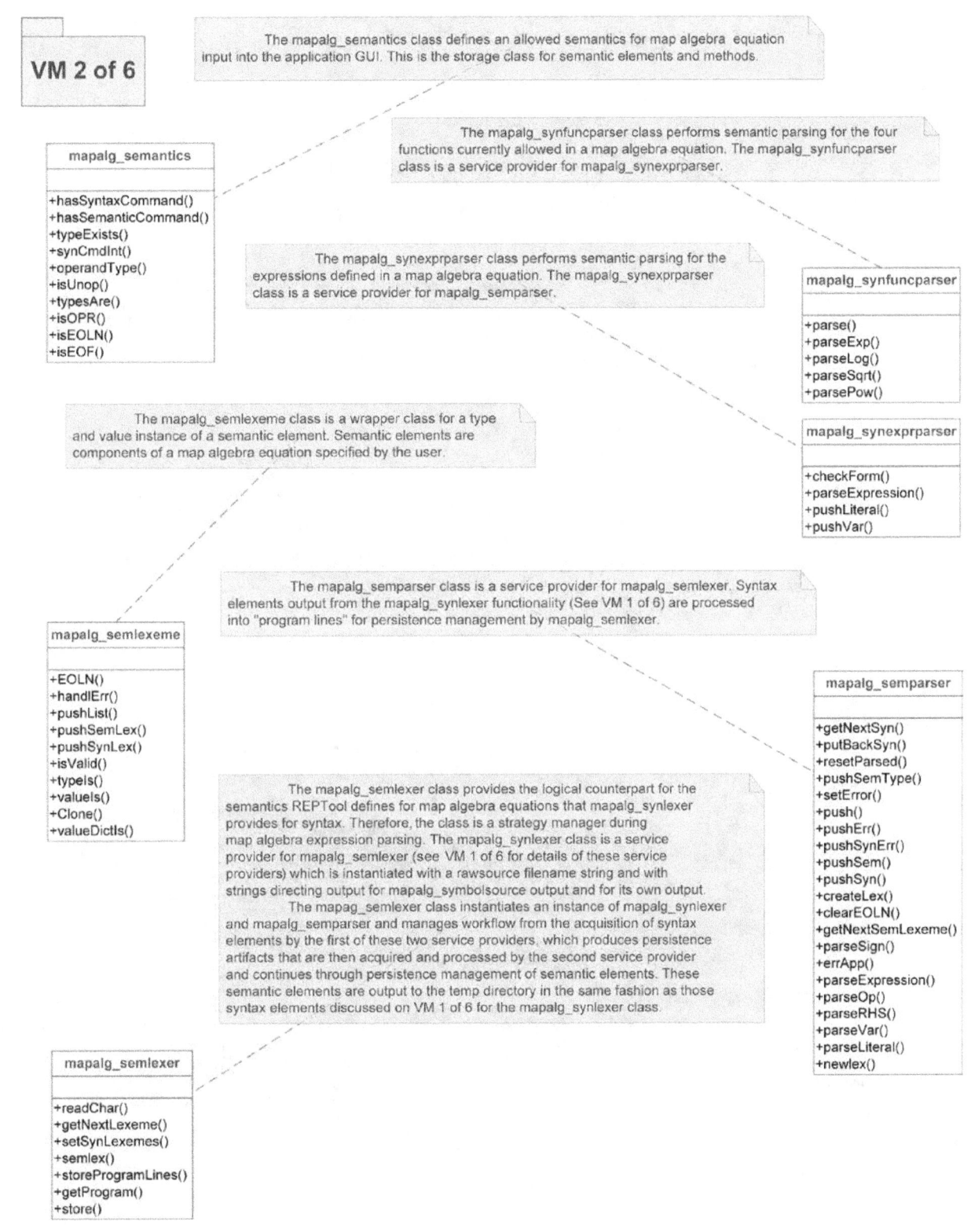

VM 2 of 6

The mapalg_semantics class defines an allowed semantics for map algebra equation input into the application GUI. This is the storage class for semantic elements and methods.

The mapalg_synfuncparser class performs semantic parsing for the four functions currently allowed in a map algebra equation. The mapalg_synfuncparser class is a service provider for mapalg_synexprparser.

The mapalg_synexprparser class performs semantic parsing for the expressions defined in a map algebra equation. The mapalg_synexprparser class is a service provider for mapalg_semparser.

The mapalg_semlexeme class is a wrapper class for a type and value instance of a semantic element. Semantic elements are components of a map algebra equation specified by the user.

The mapalg_semparser class is a service provider for mapalg_semlexer. Syntax elements output from the mapalg_synlexer functionality (See VM 1 of 6) are processed into "program lines" for persistence management by mapalg_semlexer.

The mapalg_semlexer class provides the logical counterpart for the semantics REPTool defines for map algebra equations that mapalg_synlexer provides for syntax. Therefore, the class is a strategy manager during map algebra expression parsing. The mapalg_synlexer class is a service provider for mapalg_semlexer (see VM 1 of 6 for details of these service providers) which is instantiated with a rawsource filename string and with strings directing output for mapalg_symbolsource output and for its own output.
The mapag_semlexer class instantiates an instance of mapalg_synlexer and mapalg_semparser and manages workflow from the acquisition of syntax elements by the first of these two service providers, which produces persistence artifacts that are then acquired and processed by the second service provider and continues through persistence management of semantic elements. These semantic elements are output to the temp directory in the same fashion as those syntax elements discussed on VM 1 of 6 for the mapalg_synlexer class.

mapalg_semantics

+hasSyntaxCommand()
+hasSemanticCommand()
+typeExists()
+synCmdInt()
+operandType()
+isUnop()
+typesAre()
+isOPR()
+isEOLN()
+isEOF()

mapalg_synfuncparser

+parse()
+parseExp()
+parseLog()
+parseSqrt()
+parsePow()

mapalg_synexprparser

+checkForm()
+parseExpression()
+pushLiteral()
+pushVar()

mapalg_semlexeme

+EOLN()
+handlErr()
+pushList()
+pushSemLex()
+pushSynLex()
+isValid()
+typeIs()
+valueIs()
+Clone()
+valueDictIs()

mapalg_semparser

+getNextSyn()
+putBackSyn()
+resetParsed()
+pushSemType()
+setError()
+push()
+pushErr()
+pushSynErr()
+pushSem()
+pushSyn()
+createLex()
+clearEOLN()
+getNextSemLexeme()
+parseSign()
+errApp()
+parseExpression()
+parseOp()
+parseRHS()
+parseVar()
+parseLiteral()
+newlex()

mapalg_semlexer

+readChar()
+getNextLexeme()
+setSynLexemes()
+semlex()
+storeProgramLines()
+getProgram()
+store()

Package Architecture: Virtual Machine (VM) 3 of 6

VM 3 of 6

mapalg_datalist

+setvar()
+setlit()
+setvarlist()
+setlitlist()
+varclone()
+litclone()
+varkeys()
+litkeys()
+Clone()

The mapalg_datalist class is a storage intermediary for compilation artifacts produced by map algebra equation processing. For example, the correctly formed or "fully qualified" map algebra equation—

$$(((c00*var00) + (e-1.5)) - (var01/pi))$$

—contains a coefficient variable name (c00), two raster input file variable names (var00 and var01), two constant variable names (e and pi), and finally a single literal string (1.5). The mapalg_datalist class stores these components as well as the "program" generated by mapalg_compiler.

mapalg_semfuncparser

+parse()
+parseExp()
+parseLog()
+parseSqrt()
+parsePow()

The mapalg_semfuncparser class is a service provider for mapalg_semexprparser in the same fashion mapalg_synfuncparser services mapalg_synexprparser, except these parsers are parsing out the semantic elements generated during map algebra expression processing (see VM 2 of 6 for functional notes).

mapalg_semexprparser

+checkForm()
+parseExpression()
+pushLiteral()
+pushVar()

The mapalg_semexprparser class is a service provider for mapalg_compiler in the same fashion mapalg_synexprparser services mapalg_semparser, except this parser and its service provider are not only parsing out the semantic elements generated during map algebra expression processing (see VM 2 of 6 for functional notes), but they are also performing verification and validation of those semantic elements for executability requirements of the virtual machine implementations. The mapalg_semexprparser class is instantiated with an instance of mapalg_datalist for processing artifact storage requirements.

mapalg_compiler

+handleError()
+processLine()
+handleExpression()
+handleEOF()
+getVariables()
+getLiterals()
+getProgram()

The mapalg_compiler class performs strategy management for the verification and validation of semantic elements parsed out during the processing of a map algebra equation. The mapalg_compiler class is instantiated with the filename strings for its own output and for the chain of service providers created by the two major service providers to the class, mapalg_semlexer and mapalg_semexprparser.

The mapalg_compiler class manages semantic element acquisition with an instance of mapalg_semlexer and, following instantiation of mapalg_semexprparser with an instance of mapalg_datalist, and after a call to mapalg_semlexer's method—getProgram()—the class is ready to perform verification and validation of a map algebra equation's semantic correctness. Lines of the "program" assembled by semantic parsing are then processed by the compiler for opening, intermediary, and closing formats of semantic expressions. For example, one or more semantic unary operator elements are allowed to precede a semantic expression to indicate sign changes, and their processing is handled by the compiler directly.

Processing of a semantic expression itself is handled by the mapalg_semexprparser service provider with a semantic function processing service provider of its own. Following all of this processing, if no errors occurred, the compiler retrieves the variables, literals, and "program" from the datalist instance provided to the expression parser earlier and manages persistence for these elements. Output artifacts of compilation reside under the temp directory created by input verification and validation (see VM 1 of 6: mapalg_synlexer notes).

Package Architecture: Virtual Machine (VM) 4 of 6

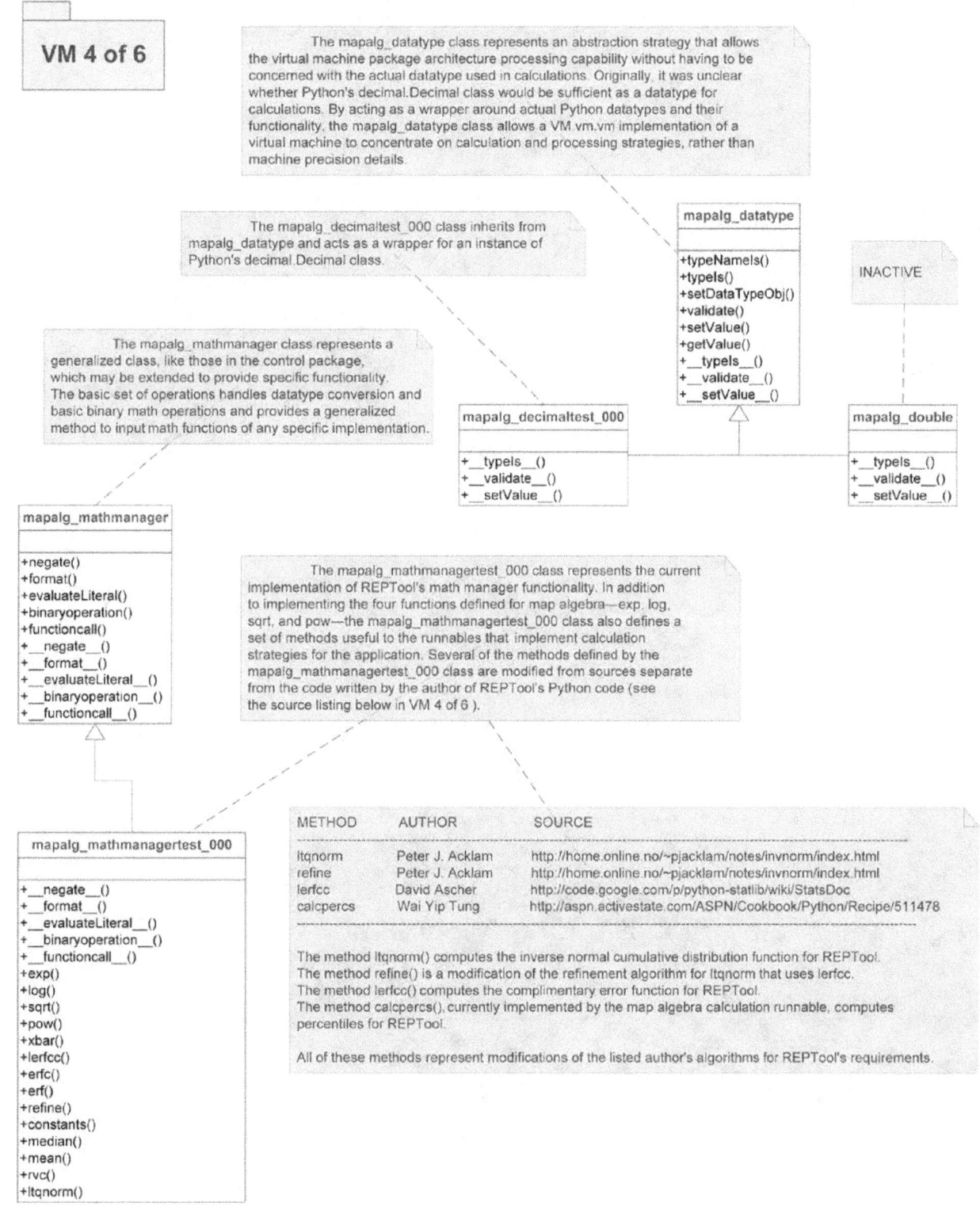

Package Architecture: Virtual Machine (VM) 5 of 6

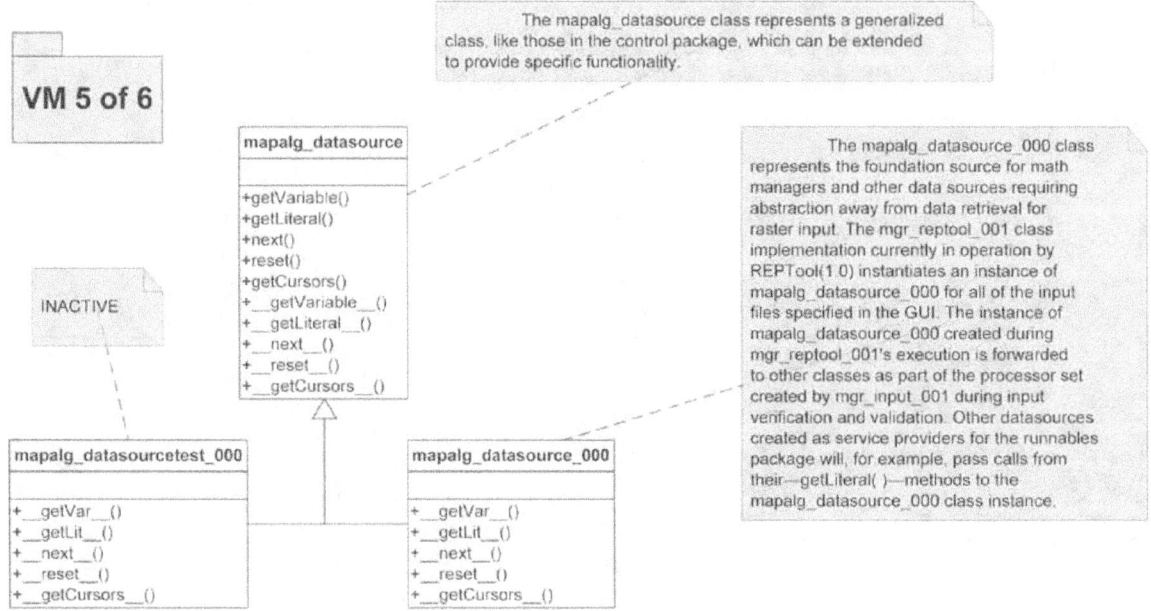

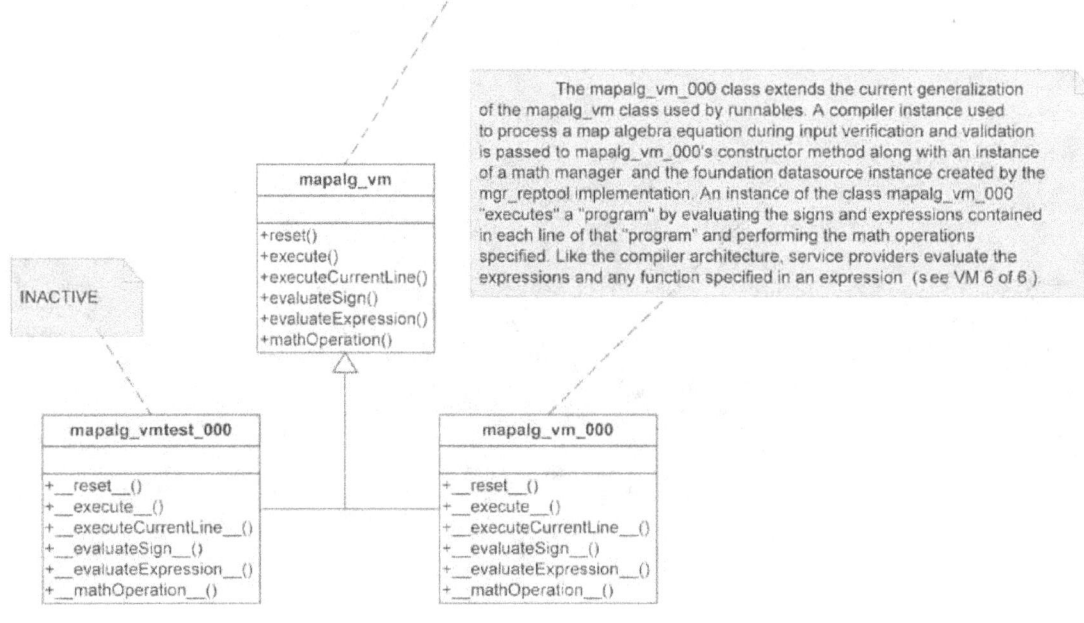

Package Architecture: Virtual Machine (VM) 6 of 6

```
VM 6 of 6
```

The mapalg_semfuncevaluator class is a service provider for mapalg_semexprevaluator. The mapalg_semfuncevaluator class handles evaluation of a REPTool defined function when the expression evaluator finds a semantically correct function specified in a map algebra equation. REPTool (1.0) semantics recognize the exp, log, sqrt, and pow functions. The mapalg_semfuncevaluator class is instantiated with a reference to the mapalg_semexprevaluator it provides services for, as well as an instance of, the datasource pulling values from input raster files and a reference to the virtual machine it ultimately provides services to. Values for variables and literal parameters specified for a function pulled from the datasource and math operations are handed off to the virtual machine for handling by its math manager (see VM 5 of 6).

```
mapalg_semfuncevaluator
----------------------------------
----------------------------------
+evaluate()
+evaluateExp()
+evaluateLog()
+evaluateSqrt()
+evaluatePow()
+evaluateLiteral()
+evaluateVariable()
```

The mapalg_semexprevaluator class is a service provider for the mapalg_vm_000 class. The evaluation of an expression consists of parsing the lexical semantic elements assembled into a "program" by the compiler implementation (see VM 3 of 6) from the map algebra equation specified by the user and performing the operations specified using a "virtual machine" (see VM 5 of 6). The mapalg_semexprevaluator class handles format verification and validation for expressions and passes off the evaluation of math functions allowed by REPTools semantics (exp, log, sqrt, and pow) to its service provider, mapalg_semfuncevaluator. Each expression contained in a map algebra equation is evaluated individually by this class with results handed back to the "virtual machine" for the overall evaluation of the map algebra equation.

```
mapalg_semexprevaluator
----------------------------------
----------------------------------
+checkForm()
+evaluateExpression()
+evaluateLiteral()
+evaluateVariable()
```

Glossary for Developer Documentation

accessor   Object-oriented programming languages may define **objects** with private or protected **attributes** (not directly accessible or visible to other objects) to protect the attributes from mishandling. However, those objects will usually define a **method** to retrieve the value stored by an attribute or even a reference to the attribute itself in a controlled manner. These methods are **accessor** methods. In Python, attributes are typically accessed directly using the *object.attribute* syntax, but accessor methods may still be a preferable option. These methods are often called the "get" methods and use the *getAttributeName( )* syntax.

attribute   An **attribute** may hold a value describing a characteristic of an **object** defined by a **class**. An automobile object has many attributes describing characteristics common to automobiles. Two of these attributes might be color and tire pressure, and the values assigned to these attributes in an **instance** of an automobile object might be blue and 35 pounds per square inch (psi), respectively. The value of an attribute may or may not be assigned to an instance of an object during execution of a program, and if it is assigned it may also be changed one or more times during the execution cycle.

class   In an object-oriented programming language a **class** defines the **attributes** and **methods** describing the characteristics and behavior of an **object**. A Python class is defined in a Python **module** file. A class defining a circle object might have attributes for radius, diameter, and circumference and methods defining calculation strategies for determining the radius, diameter, and circumference.

constructor   A **class** method which **initializes** an **instance** of the class. Python's default **constructor** for classes is the __init__( ) method. **Instantiation** of a class named SomeClass might appear as in the following line of code:

    some_object = SomeClass( )

And the __init__( ) constructor method would be called with no arguments. Following this assignment statement, the variable name, some_object , references an instance of the class SomeClass.

expression   A binary operation consisting of a left-hand-side operand and right-hand-side operand separated by a binary operator and enclosed by parenthesis delimiters. Zero or more unary operators indicating sign changes may precede either operand as well as the binary operation. Operands must conform to syntax and semantics that REPTool (1.0) defines for Map Algebra equations; they must be valid variables, constants, functions, or another expression with valid syntax and semantics. For example, in the expression

    [-(exp(4) * var00) + var01]

there is an *inner* **expression** serving as the left-hand-side operand, [(exp(4) * var00)], and it is preceded by a single unary operator indicating a sign change. The right-hand-side operand, [var01], and the *outer* expression are both preceded by zero unary operators. The left-hand-side and the right-hand-side operands of the *outer* expression are separated by a binary operator [+]. The left-hand-side operand of the *inner* expression, [exp(4)], a function defined by REPTool with the correct syntax and semantics, is separated from the right-hand-side operand of the *inner* expression, [var00], also by a binary operator, [*], and both of these operands are preceded by zero unary operators indicating sign changes. The *inner* and the *outer* binary operations are enclosed by parenthesis delimiters.

functionality   The questions *what* and *how* as they are applied to a system describe its **functionality**. A software system's services or behaviors describe *what* it does, and the specific **implementations** of code defining system **objects** and **methods** describe *how* the system performs those services or achieves its behavior; therefore, its functionality is the sum of these descriptions.

generalization   See **generalized**.

generalized   Describes a **class** definition that specifies a common set of **method** and(or) **attributes** for the general case of an **object** category. For example, a car object is the **generalized** case for sports car objects and a sports car object is in turn the generalized case for a *convertible* sports car object. A car object cannot serve as the generalized case for a jet airplane object. However, a *transportation* object could serve as generalized case for both a car object and an airplane object. In an object-oriented language an **inheritance hierarchy** typically establishes inheritance between levels of **generalization**.

graphical user interface (GUI)   A Windows dialogue screen containing user-input components such as check boxes, text fields, buttons, and file browsers.

implementation   Is the code that defines a **method, class,** or **module.** The **implementation** of a method, defined by a class, determines the behavior of **instances** of that class. A class that **inherits** a method from its **inheritance hierarchy** may use the inherited method as is, or the class may define its own implementation for that method to provide custom behavior.

inherits   See **inheritance.**

inheritance   The sharing and reuse of common **attributes** and **methods** from an **implementation** of a general category of **objects** by specific types of objects that customize the general case defines an **inheritance hierarchy**. For example, a sports car object and a delivery van object are customized cases of the car category that might have an attribute for tire pressure and a method for acceleration that are common to the car category. In an object-oriented language, an inheritance hierarchy governs the manner in which general case implementations of a **class** may be customized with additional **functionality** or modifications to the functionality inherited from the general case class. The custom case extends the general case and in turn may be extended even further, thereby providing a hierarchy of **generalization** for the objects in question. Continuing with the car example, a European-made sports car object might extend the sports car object and define a custom method for braking and, likewise, a family van object might extend the delivery van object and define a custom method for powered rear-door locks, and both of these objects would inherit the attributes and methods of the objects they extend as well as those from the most general case object, the car object, which those objects inherit from. In object-oriented programming languages like Python and Java, single (Java) or multiple (Python) inheritance is allowed; meaning classes may inherit from a single class or from multiple class definitions, respectively.

initialization   See **initialized.**

initialize   See **initialized.**

initialized   Immediately following the first occurrence in the execution of a software program where an **attribute** is assigned a value, the **state** of the attribute is considered as **initialized**. The attribute remains **initialized** until the program changes the value held by the attribute with an assignment statement. Attributes may not be used by a program until they have been initialized by an assignment statement—these attributes are considered as uninitialized, and attempts by a program to use them will generally cause an error.

instance   An occurrence of a **class** whose **attributes** have been assigned values determining the **state** of the occurrence is an **instance** of that class. For example, a class named BallClass that models a ball **object** might have a floating point number attribute for size and a string attribute for color. A program using the BallClass could then create an instance of the BallClass with the values 1.5 and "blue" for its size and color attributes, respectively, and another instance of the BallClass could be created with its attributes set to the values 5.0 and "green" in the same manner. Depending on the **implementation** of the **constructor** method for BallClass, it is possible that an instance of BallClass could be created without assigning specific values to its size and color attributes, and in this case the state of the instance would be considered **uninitialized**. For the BallClass example provided above, the size and color attributes are both instance attributes, but it is possible for a class to define class attributes as well and class attributes are shared by all instances of that class. Therefore, this means that a change in the value of a class attribute changes the value for all instances of the class.

Instantiation   Is the act of creating an **instance** of an **object** that is defined by a **class**. **Instantiation** typically takes place in an assignment statement such as the following:

```
some_attribute = SomeClass( )
```

where SomeClass is the name of a class and the syntax demonstrated automatically calls the **constructor** method for the named class. In Python, the __init__( ) **method** is used for construction, and during instantiation a class will typically have defined its constructor to **initialize** the values of the instance under construction to some reasonable initial or default **state**. For example, a set of integer **attributes** used as counters might be set to an initial value of 0, or a color attribute might be assigned the value of a background color by default to prevent accidental interference with a color scheme.

lexeme   Is the lowest-level syntactic or semantic unit of a programming language. A token (category of a language's lexemes) list representing a few examples of syntax **lexemes** defined by REPTool's programming language for Map Algebra **expressions** has the following entries

Lexemes	Tokens	REPTool Notation
c00	coefficient identifier	<var>
*	binary operator	<op>
1.5	numeric literal	<literal>
(	parenthesis delimiter	<(>
var01	raster data identifier	<var>
,	comma delimiter	<,>
-	unary operator	<->
e	constant identifier	<var>

A similar list containing a few examples of semantic **lexemes** that are constructed using sets of syntax and semantic **lexemes** has the following entries:

Lexemes Set			Tokens
<UNOP>	<->	-	unary operator
<BINOP>	<op>	*	binary operator
<BINOP>	<op>	-	binary operator
<OPERAND>	<var>	c00	binary operation coefficient operand
<DELIM>	<(>	)	parenthesis delimiter
<SIGN>	<op>	+	sign operator
<SIGN>	<UNOP>	<->	-sign operator

where the uppercase semantic notation is inside <>, the lowercase syntax notation is inside <>, and the following elements "belong to" preceding elements. For example, the *sixth* entry in the list is a semantic lexeme notated by the <SIGN> element, and the value of the semantic lexeme is defined by a set of syntax lexemes. In this example, a set of one and the syntax lexeme notated by the <op> element has the value notated by the + element; so the + element "belongs to" the syntax element and the syntax element "belongs to" the semantic element. Notice the value of the syntax element in the *first* and the *third* entry semantic lexeme happen to be the same—this is because the '-' character may assume two roles in a Map Algebra expression. Separating two operands, the '-' character acts as a binary operator, but the '-' character may also indicate a sign change if it precedes an expression or an operand. Also, note how the *final* entry defines a semantic lexeme that is constructed with a set of lexemes, and in this case is a set of one, using semantic units. The combination of semantic and syntax elements, tab separated in the example case, are line separated during actual parsing of Map Algebra input, and it is the entire chain of elements parsed out that defines an expression accordingly with REPTool's defined syntax and semantics.

method   Is a software model defined by a programming language to provide an **object** with specific behavior by **implementation** of a strategy, a calculation, or a set of instructions. A **method** may receive input and generate output, define **attributes** for its own use, or reference the attributes of objects and is generally associated with the software model where it is defined. In Java this is typically a **class** model, but other languages like Python—where methods may be defined for a **module** or a class—prove that a class definition is not the only case where a method is defined and associated with a software model.

module   A software model used by the Python programming language to define **attributes, methods**, and **classes** in an object-oriented manner. A Python **module** is stored as a file with a specific dot extension (.py) under a Python **package** folder in a file system and contains the code defining any attributes, methods, and class definitions associated with the module. Additionally, a module may contain multiple definitions (attribute, method, and class definitions).

object   Is a software model of something that exists in reality at a physical or conceptual level. A baseball, a glass of water, or an automobile are all examples of physical realities that might be modeled in software and, like each of these physical objects, an **object** has associated characteristics and behaviors. The characteristics and behaviors of an object are defined by many object-oriented programming languages (Java, Python, C++, and so forth) using a **class** to assign **attributes** (characteristics) and **methods** (behaviors) to an object.

package   Is a software model used by object-oriented programming languages, such as Java and Python, to organize files and determine scope for **objects** modeled by the language. In a file system, a **package** is simply a folder with the name of the package containing the files defining objects in the syntax and semantics of the language. In Java, package folders store files with a specific dot extension (.java) that contain Java code for Java **class** definitions. In Python, however, package folders store files with a specific dot extension (.py) containing Python code for Python **module** definitions and the file is considered the module. A Python module may contain many Python **class** definitions or it may have none, in which case, the **attributes** and **methods** defined in the module file exist independently of an object.

program   Is the set of **expressions** which conform to the syntax and semantics REPTool defines for Map Algebra expressions. A mathematical expression which has been translated into the format required by REPTool is parsed and compiled into a set of REPTool semantic elements or **expressions** that the virtual machine **implementations** can then execute. A simple example of a mathematical expression and its translation into REPTool format is as follows:

    var00 + var01

is translated to:

    (var00 + var01)

REPTool expressions consist of binary operations surrounded by parenthesis delimiters. Signs are allowed for expressions and operands inside an expression. For example, the above expression in the definition of **program** could also be written as:

    - (-var00 + -var01)

processor set   The REPTool application creates a Python dictionary during input **verification and validation** to hold references to **objects** used throughout the execution of the tool. This dictionary is forwarded to the various processing strategies defined by the application, where references are accessed, modified, and when necessary added to the dictionary. The term **processor set** refers to the Python dictionary described by this definition of processor set.

state   Is the value(s) assigned to or held by an **attribute, method**, or an **object** at any given moment during a software program's execution. For example, any changes in values held by an attribute, method, or an object change the state. The state of any attribute or object is considered to be **uninitialized** before the first assignment of a value or values occurs.

validation   See **verification and validation**.

verification   See **verification and validation**.

verification and validation   The terms **verification** and **validation** are applied to software engineering requirements under the category of quality control. **Verification** is the checklist for a requirement. **Verification** asks the question—does it exist? **Validation** is concerned more with the quality of a requirement. **Validation** asks the question—is it correct? The **verification and validation** are rarely addressed separately.

uninitialized   During the execution of a software program **state** of an **attribute** or **object** is considered **uninitialized** until it has been **initialized**. For an attribute, simply giving it a value such as 1.5 or "blue" achieves initialization, but the state of an object typically is not considered initialized until *all* of its attributes and behaviors have been assigned initial or default values.

Unified Modeling Language (UML)   Is an International Organization for Standardization (ISO) standard notation or language for designing and documenting a system in an object-oriented manner.

wrapper   The well-documented **wrapper** design pattern, also known as an adapter, solves the conversion problem between the **implementation** reality of a service **class** and the reality of a client class with an incompatible implementation.